MEASLES
VIRUS
AND ITS
BIOLOGY

EXPERIMENTAL VIROLOGY

Series Editors:

T. W. TINSLEY

Director, Natural Environment Research Council
Unit of Invertebrate Virology, Oxford, England

and

F. BROWN

Head of Biochemistry Department,
Animal Virus Research Institute, Pirbright, Surrey, England

Measles Virus and Its Biology   *K. B. Fraser and S. J. Martin*

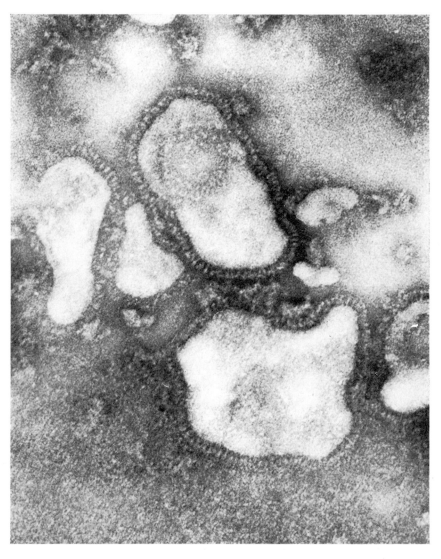

Measles virions from clarified infected vero cell extract. The particles are pleomorphic, heterogeneous in size and bear prominent spikes 12.5 nm in length.

Preparation stained with 2% phosphotungstic acid photographed on a Corinth 500 electron microscope. Magnification ×240,000.

MEASLES VIRUS AND ITS BIOLOGY

K. B. FRASER
and
S. J. MARTIN

Department of Microbiology and Immunology
and Department of Biochemistry
The Queen's University of Belfast, Belfast, N. Ireland.

 1978

ACADEMIC PRESS

LONDON   NEW YORK   SAN FRANCISCO

A Subsidiary of Harcourt Brace Jovanovich, Publishers

ACADEMIC PRESS INC. (LONDON) LTD.
24/28 Oval Road
London NW1

United States Edition published by
ACADEMIC PRESS INC.
111 Fifth Avenue
New York, New York 10003

Library of Congress Catalog Card Number: 77-74369
ISBN: 0-12-265350-5

Printed in Great Britain by
The Whitefriars Press Ltd., London and Tonbridge

PREFACE

This short monograph is not a compendium of all the published work on measles virus since it was last reviewed *in extenso*. It was written rather to acquaint the virologist with biological and clinical problems posed by this ubiquitous infection and equally to inform the clinical research worker about the behaviour of the virus which is one root of the phenomena that he sees in measles, whilst human biology is the other.

In choosing facts and in referring to published work we have been influenced by two beliefs which we hold to be true. The first is that, viruses being sub-cellular entities, the ultimate explanation of any virological process, be it *in vitro* or *in vivo*, will be biochemical or molecular-biological. The second is that, if there is any subject from which the study of virology can be divorced only by putting achievement in peril, that subject is immunology. Who would have supposed, only a few years ago, that the difference between acute and chronic infection of a susceptible cell might lie in an intrinsic defect of virus multiplication? Again, who would think that there can still be doubt as to whether immune processes cause chronic measles encephalitis, promote its establishment or simply fail to cure it?

It seems to us therefore that measles virus is a most apposite example to fit within the framework of our two assertions. If the old facts and the new which we have gathered here do not so convince the reader, then the fault lies in poverty of authorship rather than lack of information.

We are both greatly indebted to all members of the staffs of our departments who do research on measles virus or who join with us in discussion and who have helped greatly with information from the work-bench and from the literature. Their individual contributions are acknowledged in illustrations, tables and references, as are those of overseas colleagues who have generously lent figures and data from their own publications. We have been greatly helped by Mr. W. D. Linton, Mr. A. Gilmore and the staff members of the Medical, Biomedical and Science libraries of the Queen's University of Belfast; by Miss C. Lyons, Mr. M. Killen, Mr. T. McLaughlin, and Mr. R. G. Wood who also helped greatly in preparing diagrams, micrographs and photographs. We are much indebted for our own electron micrographs to Dr. Evelyn Dermott. We acknowledge most gratefully the help of our typists, Miss E. G. Halliday, and particularly Mrs. A. Marshall who had the repetitious burden of arranging and completing the main typescript.

The Queens' University of Belfast   K. B. FRASER
October, 1977   S. J. MARTIN

CONTENTS

1 | The biology of measles

INTERPRETATION—*Two principal problems, if solved, could explain many of the complications of measles. These are: the mechanisms which produce persistent non-cytocidal infection by measles virus, and the relationships between measles virus and the immune system. Whether these factors are themselves interdependent* in vivo *is not yet known.*

There have been three developments which have aroused more interest in the nature of measles virus than has been shown since the virus was first isolated by Enders and Peebles in 1954 and then adapted to make a successful vaccine (Katz and Enders, 1959; Katz *et al.*, 1962; Katz, 1965).

The most powerful stimulus was the realization that measles virus antigen was implicated in subacute sclerosing panencephalitis (Symposium, 1968), a rare but lethal disease of man, which was followed by the isolation of measles-like viruses from infected brain (Reviews: Connolly, 1972; ter Meulen *et al.*, 1972b). The reasons why a very small proportion of children, previously infected with measles, should develop a fatal, chronic measles infection of the brain years later remain unknown and will be solved only by studying the biology of measles virus in relation to immunological and other factors in the patient (Chapter 9).

The second development was unexpected. Killed measles vaccine, which was considered safer than live attenuated vaccine by many workers, not only failed to give complete protection against measles, but also sensitized a certain number of those vaccinated in such a way that subsequent natural infection caused aberrant and sometimes serious symptoms. Of these symptoms, occasional severe pyrexia, the odd change in the distribution of the rash, the onset of respiratory symptoms and residual pulmonary lesions are not only of great intrinsic interest, but are quite sufficient to have stopped the use of killed measles virus with adjuvants as vaccine until some less provocative preparation can be prepared (Chapter 13).

The third development is part of a new understanding that another chronic nervous disease, multiple sclerosis (which is much commoner than subacute sclerosing panencephalitis), has many features that suggest immunological disarrangement as a cause or consequence of the illness. Amongst these abnormalities, slightly altered and paradoxical specific reactions towards measles virus are characteristic. They consist of greater than average antibody titre towards measles virus along with slightly reduced cellular responses (Chapter 8). Nevertheless, though slight, they deserve to be investigated and that, in turn, requires more knowledge of measles virus than we have at present (Chapter 11).

Another reason for being dissatisfied with our knowledge of measles virus may or may not be intrinsic to the virus, for it arises from a problem which exists only in undernourished populations. This is a high mortality associated with lack of food, especially protein starvation (Table 1.1). In these circumstances the disease is as severe as it was 150 years ago in Western European countries which are now prosperous and relatively unaffected by mortality due to measles. It is insufficient

TABLE 1.1

The mortality due to measles in tropical Africa

Author	Place	Class of patient	Case mortality percent.
Morley (1962)	Nigeria	Hospital admission	22
McGregor (1964)	The Gambia	Village, domestic	14.01–15.22
Savage (1967)	Zambia	Hospital admission	30% (at peak, age 6–18 months).
Morley *et al.* (1967)	East Africa	In-patients	5.7%

to say that this behaviour is solely a matter of undernourishment, for few virus infections have this particular effect. Working out the mechanism behind this pathogenic process will be of the greatest interest and consequence for its cure and prevention.

In writing this account of the biology of measles virus we have been indebted to several excellent reviews, particularly those of Enders *et al.* (1957), Enders (1962), Enders-Ruckle (1963), Karzon (1962), Mc-Carthy (1959, 1962), Waterson (1965), Arakawa (1964) and Matumoto (1966). It has been impossible to avoid repeating much that has been written, but in so doing, we have tried to select properties and phenomena which are not yet completely described or understood in themselves and also those which are most likely to have some bearing on the clinical and epidemiological features of measles virus to which we have alluded.

THE EPIDEMIOLOGY OF MEASLES

The natural history of measles virus is reflected in the epidemiology of the disease. Because there is no natural reservoir of infection but man, the changing incidence of measles in a community, which we call its epidemiology, is entirely governed by the immune status of the population, which in turn is conditioned by size, density, composition by age and by previous experience of the disease. The spread of infection and periodicity of epidemics has been successfully explained mathematically (Soper, 1929; Greenwood, 1935; McKendrick, 1940; Black, 1966) in accordance with the size of a population and its previous immunological experience of measles (Figs. 1.1a,b). Thus measles tends to die out completely in communities of less than about 250,000 or so, according to birth rate, until re-introduced from without, when the great infectivity of the virus ensures its successful lodgement again.

Nothing is known of what accounts for the excessive infectivity of measles virus but it almost certainly implies rapid multiplication at the mucous surface on which it first alights whether the victim is susceptible or immune. Symptoms and signs of invasion are indeed known to occur at this stage of infection. Serological studies suggest that mean titres of antibody to the virus do decline somewhat over the years (Black and Rosen, 1962), but more so in the absence of endemic or epidemic measles (Brown *et al.*, 1969), so that sub-clinical infections can be inferred to take place. On the other hand systemic invasion by the virus as judged by second attacks, which are rare, is not likely to recur. The inference for subacute sclerosing panencephalitis is that the condition is a long standing or re-activated quiescent infection of the nervous

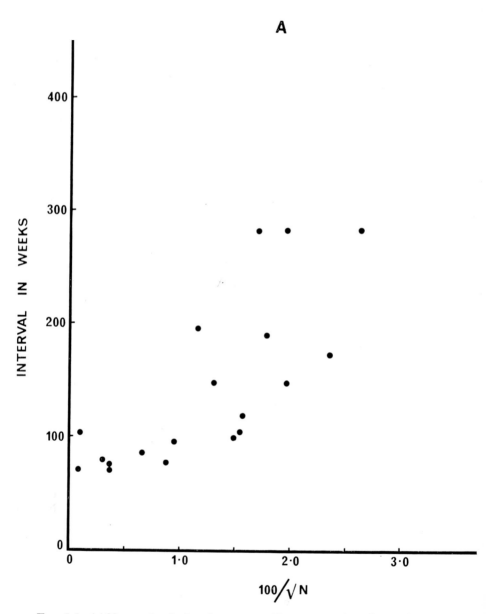

Fɪɢ. 1.1. (a)Observed relation between epidemic period and population in cities (U.K.). (Figure from Bartlett, 1957, with permission). (b) Relation between duration of epidemic and population density in islands. (Figure from Black, 1966, with permission).

B

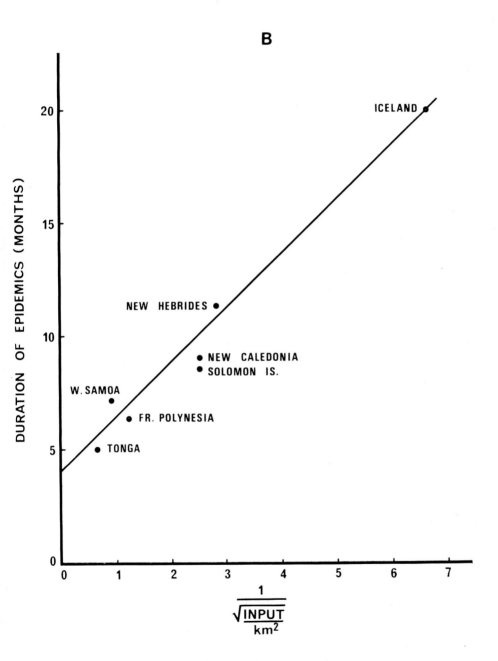

system and not a superinfection arising during a sub-clinical second attack of measles. Such latent infections have been described *in vitro* (Rustigian, 1962 and Chapter 7) and *in vivo* (Ruckle, 1958 and Chapter 9).

There is a more indirect effect of epidemiology which must be taken into account. Different communities and different countries have different modal ages for the peak incidence of measles, these being at an earlier stage of life in tropical countries (Morley, 1962) than in temperate ones and earlier in crowded urban communities than in scattered rural settlements (Table 1.2). This and other environmental conditions, such

TABLE 1.2

Age at which measles is acquired in various climates (Morley, 1969)

Place	Median age in months
N. Nigeria	16.5
Jordan	18.0
E. Nigeria	21.5
Ghana	24.7
Zambia, Rhodesia, S. Africa	29.7
England and Wales	51.7

as the incidence of breast-feeding, may well influence the patient's acquired resistance to measles as it does other infectious diseases, and so affect the incidence of complications or sequelae of infection by measles virus (Brody and Detels, 1970).

THE PATHOGENESIS OF MEASLES

If the epidemiology of measles virus is regarded as the natural history of the virus in its passage from host to host, pathogenesis represents the natural history of the virus within the host and its passage from cell to cell. Our purpose is to locate the cells and tissues which measles virus may invade and to outline the route by which it reaches them so that the lifecycle of the virus, and its incidental effects on the body, may be understood.

Stages of infection

The general picture is probably analogous to that of smallpox, as was shrewdly assessed by Grist (Grist, 1950) who constructed, from clinical and pathological knowledge, a theoretical model which followed the stages discovered by Fenner studying the pathogenesis of ectromelia virus in the mouse. These stages are primary invasion, multiplication in the reticulo-endothelial system, secondary viraemia and florid disease, after which the virus is normally eliminated (Fenner, 1948).

The route of invasion is almost certainly respiratory and symptoms presumed to represent primary invasions have occasionally been described (Goodall, 1925; M.R.C. Report, 1938). In addition, a record exists of the prevention of natural infection by the use of goggles or by the instillation of measles antibody into the conjunctival sac (Papp, 1956). This would seem to be the only experimental evidence which identifies a natural portal of entry for measles virus.

Thereafter, occasional post-mortem examination of early measles infection, supported by examination of patients' blood and also experiments in monkeys, revealed either the lesions which were deemed to be characteristic of measles virus, such as giant cells and cytoplasmic inclusion bodies, or intra-cellular virus antigen traced by immunological methods. The recovery of infectious virus by various techniques of culture occasionally added to the cytological information.

The most striking feature of such investigations is that, despite wide dissemination of virus to skin, respiratory system, intestinal tract, nervous system and renal tract, there is a strong predilection for lymphoid tissues in the early, as well as in late, stages of the disease. These tissues provide both a multiplication site and a means of transportation within the body.

Dissemination

Live virus was found to be associated with circulating leukocytes forty years ago (Papp, 1937) and the relationship was shown more precisely after methods of virus culture became available (Peebles, 1967). Experimentally, the leukopoenia* which is characteristic of the height of the disease in man, is found also in experimentally infected monkeys in which a steady drop in white cell count is most marked about the 9–11th day after infection (Taniguchi *et al.*, 1954b; Sergiev *et al.*, 1960, 1966). The reduction affects different kinds of white blood cells and it is difficult to say whether or not the count of polymorphs and lymphocytes declines simultaneously, but quantitatively there is often a relative lymphocytosis and it is characteristic of monkeys and also man that eosinophil polymorphs disappear temporarily from the circulating blood. The decrease in the number of polymorphs, in which virus does not grow, is surprising. Much infiltration of lung by

* For early references and detailed differential counts see Benjamin and Ward (1932).

lymphocytes occurs (Taniguchi *et al.*, 1954b). Measles virus grows well in mononuclear leukocytes and in both 'B' and 'T' lymphocytes in which it can establish a persistent infection *in vitro* (Joseph *et al.*, 1975b).

In fatal cases of measles the respiratory tract is the seat of heavy infection (Denton, 1925) and again perivascular leukocytic infiltration is prominent, but, at least in monkeys, the order in which lesions appear suggests strongly that proliferation and spread in lymphatic tissue precedes lodgement in skin and in the pulmonary mucous membrane (Yamanouchi *et al.*, 1970, 1973). Both these areas show signs of local infection in the way of multinucleated cells, inclusion bodies and measles antigen (Denton, 1925; Suringa *et al.*, 1970; Ackerman and Suringa, 1971).

Infection in the upper respiratory epithelium is also indicated by the presence, during the catarrhal stage of the illness, of typical multinucleate cells (Mottet and Szanton, 1961).

Direct involvement of blood vessels in the pathogenic process is suspected from the evidence of mitotic figures in capillary endothelium (Denton, 1925) and also by the recovery of measles virus antigen from vessel walls (Smith *et al.*, 1974). Yet lesions exist in the affected organs without signs of abnormality in local blood vessels (Warthin, 1931). The route through the wall, which virus must take to account for the perivascular inflammatory response, remains undiscovered.

Neurological complications of measles

One complication of measles has always been accepted as viral in origin as distinct from the commoner secondary bacterial infection of the upper and lower respiratory tracts. This is encephalitis which is said to occur in about 1/1000 patients but which is distinctly more common in older patients (Ehrengut, 1965, Table 1.3). Onset of the illness generally follows some days after the skin eruption (Ojala, 1947) and since at this time it has become difficult or impossible to isolate virus from circulating blood and also because attempts to isolate virus from CSF or post-mortem material have been equally unrewarding (Ruckle and Rogers, 1957; McLean *et al.*, 1966), there has grown up a belief that post-infectious encephalitis is not primarily the result of tissue destruction by the virus but of inflammatory, that is to say immunological, events not directly connected with *in situ* virus infection. Isolation of measles virus during and after encephalitis has recently been accomplished (McLean *et al.*, 1966; ter Meulen *et al.*, 1972c).

It is much more likely that the pathological changes are the result of virus invasion of the central nervous system and that several forms of encephalitis exist. One demyelinating variety of encephalitis is found

TABLE 1.3

Calculated approximate incidence of encephalitis per 1000 cases of measles at various ages

Reference	0–4 yrs	5–19 yrs	10 yrs	15 yrs
Greenberg *et al.* (1955)	1.01	1.60	2.98	1.67
Miller (1964)	0.68	1.1	3.8	5.3
Ehrengut[a] (1965)	0.07	0.18	0.39	x

[a] Ehrengut's collated data include those of Greenberg *et al.* (1955). x = not done.

after measles and after vaccination and is not considered here. What is surprising, in view of the wide dissemination of virus in the body, is the infrequency with which serious signs and symptoms of nervous disease appear. Pleocytosis is found in the cerebrospinal fluid of those with uncomplicated measles (Ojala, 1947) and electroencephalographic evidence certainly indicated that symptomless cerebral dysfunction is more frequent during measles than during mumps, zoster or rubella (Gibbs *et al.*, 1959). Dayan and Stokes (1971) have shown measles antigen in cells in the cerebrospinal fluid during the course of the disease.

The importance to us of actual invasion is to establish that it often occurs during a primary attack of measles. The onset of the complication subacute sclerosing panencephalitis could then be a recrudescence of an uncured infection of the brain, rather than a second infection in a previously sensitized individual. This particular and rare sequel of measles is dealt with in Chapter 9.

The part played by immunological processes in "ordinary" post-infective measles encephalitis was emphasized by Koprowski (1962) a good many years before the aetiological relationship of measles virus to subacute sclerosing encephalitis had been discovered. If electroencephalographs taken during measles are any guide to invasion of the cerebrum by the virus, then such an invasion is common, and mere penetration of the blood brain barrier is not enough to produce encephalitis. Some form of hyperergy, postulated by Koprowski as a factor in pathogenesis, may then precipitate events. Such an estimation of the part played by immune processes is also in accordance with a belief that the exanthem is a consequence of the immunological reactions between virus at the site of the rash (Suringa *et al.*, 1970; Ackerman and Suringa, 1971) and the specifically sensitized cells of the immune system. Certainly, in those patients who die of measles pneumonia because their immunological defences are naturally or artificially rendered inadequate, a measles rash is often absent (Enders *et al.*, 1959; Mitus *et al.*, 1965).

None of the foregoing evidence precludes a wholly "autoimmune" process as the cause of cerebral changes after measles, but so far the simple case of cross reaction between measles virus and the immunologically segregated normal brain has not been discovered.

Rare complications of measles

Some infrequent effects of measles point, on the one hand, to properties of the virus, and, on the other hand, to the condition of the patients in whom immunological function is often incomplete or depressed. Nevertheless, the consequences of infection are characteristic of measles, not of other viruses.

Giant cell pneumonia had been known as a separate disease from measles (Hecht, 1910; Karsner and Meyers, 1913) and then speculated upon as an infection related to dog distemper and measles by Pinkerton *et al.* (1945). Soon after culture of the virus became possible, the relationship between measles and giant cell pneumonia was first made clear (Enders *et al.*, 1959; Mitus *et al.*, 1959). Suppression of antibody formation was noted in the patients and measles virus was isolated from affected pulmonary tissue. Since then the association between immunological dysfunction due to malignant disease and generalized invasion by measles virus has been confirmed (Mitus *et al.*, 1965; Lipsey *et al.*, 1967); attenuated strains of virus are not exempt from this association, when lung, liver, bone marrow, lymph node and thymus may be heavily invaded by virus before death (Mawhinney *et al.*, 1971, and unpublished case).

Second attacks of measles are very rare (Häntzschel, 1940) and are not necessarily related to immunological deficiency, although one series of recurrent attacks in a family has been associated with other recurrent infections (Thamdrup, 1952). Such a family would bear re-investigation now that virological and immunological techniques are to hand, but Thamdrup was able to show that globulin levels in serum were normal as were responses to bacterial toxins.

The kidney provides a means of excreting measles virus complexed with antibody or enclosed in phagocytes (Llanes-Rodas and Liu, 1966), but exfoliative infection of the urinary tract has only been described as a complication of the generalized invasion seen in giant cell pneumonia (Lipsey and Bolande, 1967). All levels of the tract, from renal tubule to urinary bladder, seem to have been involved.

A beneficial effect of measles has been noted by several observers in patients suffering from nephrosis. Many of them experience short or even prolonged remissions of the condition (Hutchins and Janeway,

1947; Huth, 1963; Blumberg and Cassady, 1947). The cause is unknown but the effect could be due to the local presence of virus in view of Llanes-Rodas' and Liu's observation.

Inflammation of the intestine is a recognized complication of measles and excretion of the virus in faeces indicates replication of the virus somewhere in the intestinal mucosa. This possibility may account for the way in which measles virus causes serious deterioration in children suffering from protein starvation, but the precise mechanisms require to be elucidated.

Sensitivity to measles, produced by immunization with killed vaccine in adjuvant is, in some senses, an artefact, but the nature of the disease results in the lung being one site of the hyper-sensitivity reactions which follow and there the lesions can be particularly long-lasting, as revealed by X-ray photography (Young et al., 1970).

Lastly, measles is one of the many respiratory infections which predispose to bacterial invasion of the respiratory tract, presumably as a consequence of local destruction of mucosa, but one related observation is its undoubted deleterious effect on tuberculous infection. Here the causative mechanism is likely to be indirect and probably immunological, for, along with some other virus infections, measles does produce a diminution of cell-mediated immunity which lasts for several weeks after the disease is ended.

COMMENT

In this outline of the pathogenesis of measles, we have come to realize that there are some interesting features of the virus. The most significant is probably the close association with leukocytes and lymphoid organs in which the virus certainly flourishes and by which it is transported widely within the body.

The virus can spread to skin and mucous surfaces in particular, but it also affects blood vessels and invades parenchymatous organs like liver and kidney, when immune processes are insufficient to check it.

It not only escapes control when immune processes are already defective or suppressed but, in growing, it also produces a depressing action on cell mediated immunity and perhaps on antibody production (Mitus et al., 1962; Mawhinney et al., 1971).

The immunological effects of measles and the ability of the virus to grow in lymphocytes and monocytes can hardly be other than intimately related properties. The ability of measles virus to persist in the brain of man and in the kidney of monkeys for long periods has not yet been related to any function connecting measles virus and lymphocytes. It will be surprising if there turns out to be none.

2 | The isolation and adaptation of measles virus

INTERPRETATION—(1) *The isolation of measles virus and its adaptation to other hosts have been accomplished by random procedures. (2) Adaptation may be rapid and easy or slow and difficult and there are slight indications that the use of intermediate hosts may sometimes be necessary. (3) Cell fusion may help to by-pass species barriers. (4) These facts point to the possibility of phenotypic as well as genotypic variations. Both kinds of variants should be searched for.*

There are adequate reasons for re-examining the two problems of isolating measles virus and of adapting it to various species of host. Firstly, the relatively greater recovery rate from infected cells as compared with that from infected fluids suggests that cell to cell transfer, and therefore cell-membrane, is likely to be involved. Secondly, persistent non-productive, intra-cellular infection by measles virus has been found to be associated with imperfect assembly at the cell membrane (p. 141) and, thirdly, some related paramyxoviruses depend for their infectivity on the action of host-derived proteases which make the virion capable of haemolysing red blood cells and of fusing cell-membranes (Choppin et al., 1975). These three observations each point to the importance of this site of activity for virus function, especially infectivity.

The question of whether adaptation involves genetic selection of the virus, acquisition of helper enzyme from the host, or both, is therefore likely to be of importance also to measles virus.

ISOLATION OF MEASLES VIRUS

Soon after the discovery that measles virus could be grown in tissue culture, McCarthy remarked on the narrow range of natural infection by this virus (man and monkeys which are kept in the vicinity of man) and compared this with the relatively wide range of cell species which could be infected in the laboratory (McCarthy, 1962). Since then further habituation has increased the number of animal hosts and cell cultures to which measles virus can be regularly adapted but in which it does not grow when introduced directly from the patient. Some important changes must occur between the virus in measles and the freshly isolated virus and again between each laboratory adaptation in order to account for alterations in host range.

The effect of measles virus on tissue culture was first seen on a culture of human embryo kidney at the 7th day after inoculation of material from throat washes or from throat swabs suspended in sterile milk. Areas of obliteration of cell boundaries and cytoplasmic vacuolation gradually assumed the structure of multinucleate "giant-cells" in which inclusions were later seen. Margination of chromatin and the formation of a central eosinophilic body in the nucleus were noted as early as the 4th day after primary inoculation of the culture and simple tests showed the difficulty of passing the virus immediately either in baby mice or in 7-day-old eggs (Enders and Peebles, 1954). In a later publication, Enders (1956) reflected that the prominent intra-nuclear inclusions made staining of virus inclusions in the cytoplasm an unexpected finding, but there was no suggestion that they were late developments of laboratory artefacts. He also reported that 8 strains of virus had been isolated from 8 patients, using blood as a source, a success rate of which the true significance was probably not realized at the time.

In the same year in which tissue culture first proved successful in isolating measles virus, Taniguchi and his colleagues produced fresh evidence which suggested very strongly that unadapted measles was transferable with some regularity from man to some species of macacus monkey (Taniguchi et al., 1954b). Enders' success with monkey kidney as well as human embryo kidney added to the similarity between the two species as hosts. Some years later this was confirmed and direct isolation of measles virus from throat washes or blood was accomplished from material taken on the first day of the rash, 5 out of 9 patients yielding virus (Bech and von Magnus, 1958). Monkey kidney cultures had already yielded some isolates of measles virus out of 13 viruses recovered from 31 patients following the inoculation of blood or faecal filtrate into human amnion and monkey kidney cultures (Frankel et al., 1957).

Unfortunately, this preliminary report revealed that some uninoculated monkey kidney cultures also grew measles virus so that the true recovery rate from patients is not known. It was important evidence that the monkey is a natural host for measles virus and probably an early pointer to the need for cell replication in releasing latent infection (see Chapter 8). Dog kidney in cultures also served for the primary isolation of measles virus from all 5 samples of clinical material in Frankel's hands (Frankel *et al.*, 1958), a point of some importance in view of the relationship of measles to dog distemper (see p. 37).

Not much significant information can be gained from any other studies of primary isolation of measles virus except to say that measles virus was isolated from patients' leukocytes on human amnion cell cultures by Gresser and Chany (1963) and in HeLa cells by Shingu and Nakagawa (1960), who also used blood as the source of virus, obtaining 3 isolated out of 5 attempted. Shingu and Nakagawa allowed 7 to 14 days of incubation for virus to develop. Gresser and Chany's rate of success was 10 isolations out of 11 attempts on the first day of the rash and only 1 out of 8 on the 2nd day; they emphasized the need to leave the culture undisturbed for several days. The use of human blood almost certainly owes its success as a source to the presence of measles virus in leukocytes.

Enders had noted that primary cell cultures make more sensitive substrates than continuous cell lines (Enders, 1962), but no formal tests have been carried out to confirm this or to explain it. Some standard strains of measles and SSPE measles viruses appear in Table 2.1.

ADAPTATION TO TISSUE CULTURE AND CHICK EMBRYO

From the time when it was first isolated measles virus has been known to multiply in various human cell cultures and in cultures of monkey kidney (Enders *et al.*, 1957). Matumoto concluded from a study of the available reports that human embryo kidney and monkey kidney were equally sensitive for the primary isolation of measles virus (Matumoto, 1966). Once this is accomplished, conversion to growth in continuous cell lines seems to be reasonably easy. Five or 10 passes in KB cells is mentioned by Kohn and Yassky (1962) and HEp_2 and HeLa cells were shown to be susceptible to measles virus at an early stage in laboratory study (Dekking and McCarthy, 1956; Black *et al.*, 1956; Adams *et al.*, 1956). In other works references occur to human amnion cells, conjunctival cells and human liver cells (Enders *et al.*, 1957; Bech, 1958; Mutai, 1959), but none of these papers deals with investigation of the

TABLE 2.1

Standard strains of measles virus[a]

Name	Date Isolated	Culture Used	Reference
Edmonston	1954	Human kidney	Enders and Peebles, 1954.
Philadelphia 26	1957	Monkey kidney	Frankel, *et al.*, 1957.
Sugiyama	1958	Monkey kidney	Matumoto *et al.*, 1959.
Leningrad	1958	Human kidney	Smorodintsev *et al.*, 1965.
Tanabe	1958	Monkey and human kidney	Ueda *et al.*, 1974.
Toyoshima	1959	Human amnion, FL cells.	Toyoshima *et al.*, 1959b.
1677 Marburg	1960	Human kidney	Ruckle-Enders, 1962
Pittsburgh	1955	Human kidney	Ruckel, 1957
Some SSPE measles strains			
SSPE (1)	1969	Biopsy culture. Co-culture Hela.	Horta-Barbosa *et al.*, 1969b
SSPE (2)	1969	Biopsy culture. Co-culture Hela.	Horta-Barbosa *et al.*, 1969b
JAC	1969	Brain biopsy culture. Co-culture CVI cells.	Katz *et al.*, 1969
LEC	1969	Brain biopsy culture. Co-culture CVI cells.	Katz *et al.*, 1969
RO-SSPE 1	1969	Brain biopsy. Co-culture with BSCI cells.	Payne *et al.*, 1969
DU–H76774	1970	Biopsy. Intra-cerebral into hamster.	Parker *et al.*, 1970

[a] Details of these strains are available in Majer (1972).

mode of adaptation from one tissue to another. Nearly always the first pass from one host line to a new species is marked by a pronounced retardation of virus reproduction or of CPE which is speedily overcome at the next one or two transfers. This rapid, almost sudden, change suggests that some host-controlled phenotypic or physiological difference between the product of one infected cell line and another, as well as the selection of genetic variants, should be considered as a mechanism of adaptation. Rapp (1960) showed by simple cloning of a population of HEp_2 cells that individual clones showed remarkable differences in their capacity to produce infectious viruses, as much as a difference of five log_{10} $TCID_{50}$. This is probably quite different from the ability to infect which, with uncoating, is the first step that requires investigation in the rapid fluctuations that we have been referring to.

The first principal adaptation to be sought outside man and monkey was to grow virus in the fertile egg for the purpose of preparing an attenuated vaccine. There was information available suggesting that growth in the chick embryo might be possible following a primary inoculation. Rake and Shaffer (1939) claimed, after 29 passes in the chick embryo by the chorio-allantoic method in embryos at 12 to 14 days old, the production of a fluid that was still infectious for monkeys, though somewhat attenuated, compared with the same material taken at an earlier stage of its history. It is probable that they also demonstrated the production of post-infectious immunity. Their starting material was bacteria-free throat wash or unfiltered blood.

Milovanovic, on the other hand, studied adaptation to the chick amniotic membrane with Edmonston virus that had already been passed 24 times in tissue culture of human embryo kidney. The virus was re-adapted to human amnion with some difficulty, only one strain out of eight being passed successfully. This strain was used, after 28 further passes in human amniotic cell line, to get egg-adapted virus in the chick amniotic cavity (Milovanovic et al., 1957). Six passes of this material in tissue culture of chick embryo resulted in an incomplete destruction of the cells, but an incubation period diminishing from 30 days to 5 days. The decay of the virus at 37°C was said to be rapid (Katz et al., 1958).

It may be of some significance that measles virus which had been prepared in human embryo kidney could not be grown in, or adapted to, chick embryo. The Toyoshima strain of measles virus was moderately adaptable to the chick embryo by the amniotic and the chorio-allantoic routes of inoculation after being carried in the FL line of human amnion cells. It is of great interest that adaptation to the two tissues in the one species of embryo resulted in loss of pathogenicity for human FL (amnion) cells by the chorio-allantoic passage and retention of it by the virus passed in the amniotic cavity (Okuno et al., 1960). Here the tissue, rather than the species, seems significant. The same was true of Edmonston virus treated in the same way. When Chiang (1966) studied the growth and distribution of virus in the 7-day-old chick embryo, he used the chick-amnion-adapted derivative of these two derivatives of

Toyoshima virus and, next to amnion itself, found most measles virus in embryonic intestine.

A remarkably early adaptation to a wide host range was described by Wright (1957a,b) who detected measles virus-specific antigen in human and rodent cell cultures, including those of mouse, guinea-pig and hamster, that had been infected with a fresh isolate grown in monkey kidney cells. A cytopathic effect was seen, but no transmission of virus was demonstrated. The unusual host range does raise the possibility of mixed infection with another virus, such as herpes simplex. However, there are reports of growing virus in guinea-pig cells. One, without detail, is given by Mascoli et al. (1959) and mentions giant cell formation in guinea-pig spleen cell in vitro. The other is an interesting study by Cernescu et al. (1973), of adaptation to guinea-pig kidney cells in which low temperature variants were recovered. There were early reports that measles virus does not grow in cultures of rabbit kidney cells (Brown, 1957; Ruckle, 1957) and this was confirmed by Mauler and Hennessen (1965) much later, but these authors did observe chromosome breakage in some of the inoculated cells.

The Edmonston virus with a history similar to that given by Milovanovic plus 8 more passes in human amnion proved adaptable to bovine kidney tissue culture in tubes. Eight blind passes resulted in a cytopathic effect appearing or, alternatively, after 81 days' incubation in bovine kidney cells, release of virus into the supernatant fluid supervened and the released virus, on passage, proved cytopathogenic in 15 days at the 4th pass, instead of 70 days at the first. This adaptation was accompanied by loss of pathogenicity for monkeys on the 20th pass, but the virus did grow in the monkey, being recovered from throat swabs and from blood (Schwarz and Zirbel, 1959). A similar adaptation to calf kidney was carried out with a Japanese strain of measles virus, Sugiyama, through monkey kidney and human conjunctival cells, with another, the Tanabe strain, through only 4 renal cultures and with Edmonston virus through human amnion, monkey kidney and FL cells. In the first strain the incubation period was 14–20 days at first, slowly decreasing to 5 days at the 25th pass, whereas the Tanabe strain required only 3 days for CPE to appear; the Edmonston strain took 5 or 6 days from the time of inoculation. The yields of virus were related to the length of the incubation periods (Matumoto et al., 1961).

Of other species, sheep kidney cells have been shown to give a long latent period after infection. These were primary cell cultures, but different strains of virus (Edmonston or Tanabe virus), grown in various cell lines (human, monkey or bovine), were equally easily grown in the culture (Makino et al., 1970) and showed optimum cytopathic effect and yield at 20 to 21 days after infection. In this model a course of adaptation is unnecessary but on passage shortening of the incubation period required for cell-destruction was observed. It seems likely that ferret kidney is another primary culture in which an immediate take follows inoculation of Edmonston virus (Shaver *et al.*, 1964). Hamster embryo fibroblast as a vehicle for latent measles infection is dealt with on p. 103 (Knight *et al.*, 1972).

Finally, interest in neurotropism and other neurological effects of measles has resulted in the testing in neural and non-neural tissue culture of measles virus strains that have been derived from infections of brain (p. 155).

Dog kidney, like bovine kidney, would seem to be a likely tissue for growing members of the measles–distemper–rinderpest group of viruses. Musser and Slater (1962) showed that Edmonston virus which had 22 passes in chick embryo proved capable of inducing cytopathic changes in kidney cell cultures derived from beagles in which they had taken a good deal of trouble to exclude common bacterial and relevant viral infections. A stellate form of cytopathic effect appeared after only 8 passes and a steady yield of virus occurred, between 10^5–10^6 ID_{50} per ml from the 6th to the 16th day after infection. This type of cytopathic effect had been observed by Frankel originally at the edges of the infected monolayer; he claimed in his original preliminary report (Frankel *et al.*, 1958) that it was possible to isolate measles virus directly in renal cultures from dogs. Neither the virus strain nor the nature of the maintenance medium seemed to affect the outcome of infection in the cultures. Similar cultures have been used by Tawara (1964, 1965) to study the morphogenesis of the virus but the process of adaptation of virus is not described.

Just as measles virus was adapted to the mouse by intra-cerebral inoculation, so it has been adapted to *in vitro* growth in mouse cell-cultures. A continuous line of mouse embryo cells, in which measles virus replicates, was used by Parfanovich *et al.* (1966) to show the effect of measles virus infection on the function of the nucleolus, but the process of adaptation and cytopathic effects were not stated. In mouse L cells, on the other hand, adaptation did not occur; following an initial increase in titre during the first 36 h after infection, the amount of virus present decreased and only the number of infectious centres remained constant for the next 70 days or so. There was no cytopathic change; uninfected cells were recovered from the culture (Kohno *et al.*, 1968). Experiments showed that penetration of the virus into the cell was rapid, but uncoating was very slow compared to what happened to the same virus in Vero cells. These papers suggest models by which modification of various strains of measles virus to growth in mouse tissue culture could be studied.

Organ culture of spinal dorsal root ganglion and of cerebellum have provided a long-term infection with Edmonston virus. Strictly speaking this was not adaptation, infection being immediately successful with a steady yield of infectious virus. Neurones are affected from the 18th day onwards (Raine et al., 1969, 1971). Explants of mouse brain were susceptible to infection by a strain derived from Edmonston virus, and infection was marked by the small proportion of cells carrying antigen. There was little or no antigen present at the cell membrane, and the yield of infectious virus was very poor (Gibson and Bell, 1972).

Thus there is a fairly wide variation in the ease with which measles virus may be adapted from tissue culture of one species to that of another. However, the effects of multiplicity of infection, action of interferon, of non-infectious virus, selection of variants and many other factors cannot be elucidated by the method of blind passage which has mostly been used.

There are two matters for regret in all the information from these papers. The first is the great number of experiments made with a single laboratory strain of measles (Edmonston) or derivative of the strain, some of which have had long and varied passage before being used. The second is the lack of information about the steps involved when virus fails to grow. There is almost no systematic study available of adsorption, penetration, uncoating, or synthesis of nucleic acid and protein by unadapted and ineffective virus. Assembly and release have been partly investigated in defective infection by neurotropic strains of measles virus (p. 141).

ADAPTATION TO MAMMALIAN HOSTS

The production of a systemic measles-like illness in animals from the inoculation of human material has been achieved in monkeys of various species (Taniguchi et al., 1954a,b; Sergiev et al., 1960; Yamanouchi, et al., 1970, 1973). Many older similar claims were probably genuinely successful (e.g. Anderson and Goldberger, 1911; Blake and Trask, 1921a,b), though not well substantiated because the varying state of immunity in the subject animals interfered with serial transmission of the disease (Enders et al., 1957). Some excellent pathological studies were done on infected monkeys before the virus had been cultured and characterized (Hathaway, 1935; Gordon and Knighton, 1941; Taniguchi et al., 1954a).

However, extension of the host range to non-simian animals proved not too difficult, though sometimes tedious, when tissue-culture-adapted measles was chosen as the starting agent. The chick embryo was the first animal host outside the monkey family to be proved susceptible and in it amniotic inoculation was the chosen route of adaptation (Milovanovic *et al.*, 1957) (p. 16). The first inocula actually consisted of measles virus-infected human amnion cells.

The unfortunate laboratory mouse was the next victim when Imagawa and Adams (1958) adapted the Edmonston virus from HeLa cells to suckling mice by the intra-cerebral route and then Waksman *et al.* (1962), having adapted the Philadelphia 26 strain of measles virus to hamster intra-cerebrally, noted that late passes of hamster neurotropic virus were also pathogenic for *infant* and *adult* mice. When Matumoto adapted the Sugiyama measles virus, which was a monkey kidney isolate originally, he made one observation which may give a clue to the requirements for adaptation. None of his attempts at passage intra-cerebrally in mice succeeded unless the virus had some history of passage in human amniotic (FL) cells. The nature of this requirement has not been investigated (Matumoto *et al.*, 1964). Successful adaptation to mouse brain from human virus was claimed first by Arakawa (1948, 1949), who found adaptation difficult as compared with the isolation of SSPE measles virus reported by Greenham *et al.* (1974).

Since then most published papers on the subject of adaptation to animals have dealt with neurotropic properties of measles virus. This is largely a consequence of interest in subacute sclerosing panencephalitis. Burnstein and Byington (1968) found that hamster-neurotropic virus, at the 109th pass in hamsters, grew in the brain of Wistar rats at less than 3 days of age, producing viruses that were infectious for hamsters by the intra-cerebral route. This rat brain virus was not able to grow in BS–C–1 cells in which the parent hamster neurotropic virus grows well, but cells from infected brain used as infectious centres on BS–C–1 cells revealed the presence of virus. A similar defect was characteristic of Imagawa's and Adams' measles virus that had been adapted to suckling mice; it would not produce extracellular infectious virus from HeLa cells, although a cytopathic effect and cell-associated virus were both present in inoculated HeLa cell cultures. Measles virus isolated from brain of patients with sub-acute sclerosing panencephalitis and passed in monkey kidney cells proved neurotropic for hamsters in early passage (Lehrich *et al.*, 1970) and this characteristic, hamster neurotropism, has been examined in various adapted strains of measles virus (Albrecht and Schumacher, 1971; Janda *et al.*, 1971; Mirchamsy *et al.*, 1972).

Apart from hamster-brain-adapted virus itself which was 100% pathogenic, Albrecht and Schumacher recorded encephalitis rates varying from 5 to nearly 40% according to the strain of measles virus, the highest being Edmonston virus which had been passed 28 times in human amniotic cells. This confirmed a solitary observation by Wear and Rapp (1970) on another derivative of Edmonston virus, the attenuated Schwarz strain. It should be noted that these strains had been pased in monkey kidney cells before being tested in mice.

Janda et al. took the precaution, when they tested 4 strains of Edmonston-derived virus, of passing them all in the same cell line at least once (1 to 4 times) before testing. They tested pathogenicity and also estimated the rate of adaptation from the decreasing length of incubation found as serial passage progressed, noting that these properties differed from each other in various strains. For example, Edmonston virus was the most pathogenic and showed rapid adaptation; virus prepared from highly diluted inocula had very low initial pathogenicity and rapid adaptation, whilst undiluted inocula had high pathogenicity but slow adaptation. Both Albrecht and Schumacher, and Janda et al. note that there can be symptomless infection, either with histological evidence of encephalitis (Albrecht and Schumacher, 1971) or showing gradually increasing yields of virus per brain on serial passage. The titre did not increase after symptoms had appeared (Janda et al., 1971).

In the study by Mirchamsy and his colleagues, an age requirement for adaptation was noted when testing attenuated viruses. Hamsters when 1 or 2 days old, provided a 100% pathogenic virus from all of four commercial vaccines by the 8th intra-cerebral pass. No illness was produced in 5-day-old animals. The lesions produced by the adapted vaccine strains differed from those produced by fully adapted "neurotropic" strains of measles virus, and the adapted vaccine strains were of different lethality. The account in this paper implies that only an occasional hamster had shown signs of encephalitis when attenuated vaccines were being tested for safety, whilst monkeys, guinea-pigs and mice were insusceptible to the same preparations. Yet formal testing showed 20 to 30% morbidity at the first intra-cerebral pass in suckling hamsters. It may be that this fairly high take is a consequence of preliminary passage of the vaccine strains in Vero cells, a measure which is known to accentuate neuropathogenicity of vaccine strains quite markedly (Mirchamsy et al., 1972). If so, it would be a point of great interest to the study of adaptation.

Japanese workers have tested one SSPE, non-yielding, virus strain (Niigita) and three other measles viruses, including two vaccine strains, for neurotropism in hamsters; the cultural histories of the strains differed widely up to the time of use. The Niigita strain required no adaptation, growing in infant and adult hamsters and producing fatal encephalitis; the others, all lethal for newborn hamsters, increased in titre for 3 days in adult animals after intra-cerebral inoculation and then declined in amount. One hamster immunized by a different measles strain (TYCSA), developed subacute meningo-encephalitis from the SSPE strain (Katow *et al.*, 1973). Katow *et al.*, testing several adapted strains in newborn hamsters and mice, noted wide differences in pathogenicity after intra-cerebral inoculation of virus. For example, four strains tested were highly virulent to newborn hamsters, but only two of those were virulent to mice, the Schwarz and Edmonston strains being non-lethal. Some strains could be recovered from hamster brain, some not. The age limit of hamster susceptibility with their strains was 9 days, not 5 days as with Mirchamsy's viruses.

None of these works is a full study of adaptation to brain tissue and none has attempted to determine requirements for adaptation or to analyse changes in the virus as adaptation proceeds. Alteration in cytopathic effects produced on W-F, human amnion cells by neuro-adapted viruses was observed by Mirchamsy and his colleagues.

Clinically silent infection of brain has been referred to (p. 125). Albrecht and Schumacher (1971) and Wear and Rapp (1971) have shown that non-neurotropic measles virus can persist in the brain of hamsters which are immune at birth (protection conferred by maternal antibody had been noted in mice by Imagawa and Adams in 1958). This chronic infection could be re-activated, but it is not stated whether adaptation of the persistently-infecting virus had taken place through long residence in hamster brain.

SSPE MEASLES VIRUSES

The brain is a specially segregated organ on which tests of susceptibility to measles virus by intra-cerebral inoculation may not be a true indication of host range in the meaning of range of species affected. For this reason the neuropathogenicity of SSPE measles viruses is dealt with in other chapters (p. 135 and p. 155). The species named in those chapters as being susceptible to SSPE measles viruses are: mouse, hamster, ferret, dog, calf, lamb and monkey. In these species intra-cerebral infection is most easily produced by slow-yielding, persistently infected cells, which implies that cell-to-cell transfer is advantageous in the brain. There is no evidence that the SSPE viruses are truly neuro-virulent mutants of measles virus.

COMMENT

No constant pattern or sequence of behaviour can be discerned in present accounts of the adaptability of measles virus to animal hosts other than monkey and man, or to other species of cell-line, nor are there any distinct indications of what to investigate by way of explanation. Amongst possibly significant observations we might consider the following:

(1) Contact between infected and non-infected cells seems to bypass some of the difficulties in infecting across species barriers or even crossing from natural host to tissue culture of the same species. The examples are, the relative ease with which measles virus may be recovered from leukocytes as opposed to fluids, provided that time is allowed for undisturbed contact between cells, and also the ease with which cell-associated SSPE virus infects animal brain, in contrast to the difficulty experienced with disrupted cells or free virus.

(2) There is great variation in the rapidity of adaptation, varying from one pass to many "blind" transfers. The difference raises the question of two kinds of adaptation, one phenotypic or host-controlled, the other conventional selection pressure for suitable mutants. There is no example of adaptation being successful during one growth cycle, but there are examples of significant change of infectivity following one transfer in a new host.

(3) There is evidence that passage in an intermediate host may be necessary either for rapid adaptation or for a more gradual change of host-range. One or two examples are seen in which passage in monkey kidney cells can convert non-neurotropic attenuated measles virus into a neurovirulent organism. Similarly, numerous passes in human amnion cells have been reported as pre-requisite to adaptation to chick embryo or to mouse. There is no evidence as to whether change of host allows the virus to carry with it from one host genetic or enzymic material which would assist its growth in another, but the possibility needs bearing in mind.

(4) It is just possible that measles virus which has been adapted to propagation in continuous lines of monkey cells is less easily adaptable to chick embryo or to infant mouse than "wild" measles virus, as witnessed by the need to use cultures of human amnion as an intermediary host. This may be connected with the same hypothetical mechanisms as are mentioned in point 3, but the existence of two sorts of haemagglutination (conventional and salt-dependent) by measles virus should be taken into account (p. 58). If the conventional sort represents tissue-culture-adapted or monkey-fixed virus, it may be that the salt-dependent virus is easier to adapt to new hosts, because it more nearly resembles the natural or starting virus. This possibility is open to examination.

(5) Recent work on paramyxoviruses has shown a close association between infectivity of the virus, haemolysis and cell-fusion on the one

part and the cleavage of virus-incorporated proteins by proteases in host cells on the other. It may be that adaptation can be phenotypic by acquisition of suitable enzymes from cells in which the virus grows or genotypic by selection of variants which will react most suitably with cleavage factors produced by one or other kind of host cell.

(6) There has not been much work done from which to assess the influence on host range of virus growth-rate or of virulence. Some papers have noted an increasing pathogenicity without increase of infectious titre of the virus, others show improving growth rates as adaptation proceeds. Most estimates of virus yield in relation to adaptation appear in studies of neuropathogenicity. We have mentioned earlier why infectivity is an unreliable guide to the growth of neuro-adapted virus.

(7) Almost nothing has been done to ascertain at which stage of the virus growth-cycle an insusceptible host is non-permissive. The reason for concentrating on the cell membrane, that is on adsorption and penetration as a definition of "infectiousness", is the evidence that cell fusion can by-pass some of the obstacles encountered by the virion on a new host. These are, therefore, most likely to be surface effects. Thus, when we consider the extraordinarily interesting diminution of "virulence" for human amnion which results from passage of amnion-adapted virus in chick chorio-allantoic membrane, when no change occurs as a result of parallel passage in the amniotic cavity of similar chick embryos, we are at a loss to know if this is deficient "infectiousness", i.e. adsorption and penetration, deficient synthetic ability, deficient maturation or simply a failure to produce a cytopathic effect. Tissue-tropism *in vivo* is analogous to host-range *in vitro* when cultures from different animal species or different tissues are examined. In this context, neuropathogenicity of SSPE viruses which can, by virtue of their particular cell-associations, bypass species barriers, may not represent so comprehensive a host range as the number of sensitive species implies.

(8) Finally, there is the earliest of all observations on adaptation of measles virus to laboratory cultures. Enders pointed out that primary cell cultures were much superior to continuous cell lines as means of isolating measles virus. If this is always so, the reason should be worth searching for.

3 | Biological characters and genetic markers

> INTERPRETATION—(1) *A few naturally occurring character differences of measles viruses, notably the formation of large syncytia, large or small plaques and of salt-dependent haemagglutinin are easy to maintain at limit dilution.* (2) *High titre stocks of unmixed character are difficult to prepare.* (3) *Isolates may also show different host range, cytopathogenicity and plaque formation which may be the result of selection during isolation.* (4) *The antigens seem to be invariable.* (5) *Stable induced temperature-sensitive and other mutants can be prepared.*

In order to be identifiable a virus must have one or more invariable characters. Influenza A, for example, has properties in common with orthomyxo-viruses B and C, but the soluble or nucleocapsid antigen is the constant feature, common to all variants of influenza A, that is quite different from the corresponding antigen of either B or C influenza viruses. Whether the nucleocapsid antigen of measles virus is common to and constant for the measles, rinderpest and distemper group of viruses or constant only for measles virus has not yet been settled, although recent work by Norrby carries the very important implication that it may be common to the group (Orvell and Norrby, 1974). In that event, measles, rinderpest and distemper viruses would be analogues, not of influenza A, B and C, but of the equine, avian, porcine and human strains of influenza A.

Be that as it may, variants of measles virus are not numerous and very few have been well characterized; they are mostly laboratory-adapted strains and there is no firm evidence that variation of the virus occurs naturally between countries or between epidemics. Serologically there is none, for attenuated vaccine has been as universally successful against measles as has vaccinia against smallpox. True mutants of a parental stock measles virus are very recent innovations, which, so far, have no counterparts in nature.

It is optimistic in the circumstances to use the term genetic marker. What we have done in this section is to reiterate certain characteristics of measles virus and to assess tentatively their probable usefulness in identifying various isolates of measles virus and their probable potential in future genetic experiments.

The order in which we consider them happens to be roughly that in which they were first observed and not the order of importance to the genetics of measles virus, knowledge of which does not yet exist. The characteristics are: host range, cytopathic effect, pathogenicity in animals, plaque-formation, haemagglutinin and its corollary effects, SSPE strains of virus, temperature-sensitive and other induced mutants, and physical characteristics.

HOST RANGE

Only a few years after measles virus had been isolated, reports began to appear about its cultural behaviour and its adaptation to other species in the laboratory (Enders et al., 1957; Enders, 1962); its stability and antigenic uniformity in nature had been noted (McCarthy, 1959, 1962) and the term "marker" had been used in comparing the characters of virulent, with attenuated, strains of the virus (Enders et al., 1962). These biological similarities and differences nearly always referred to derivatives of the original Edmonston strain of measles virus and there was nothing to indicate that naturally occurring varieties of measles virus existed. This aspect of the situation has altered little since then.

Host range changes by adaptation in the laboratory and the fact is established that vaccine strains, attenuated by passage in the chick embryo, lose their virulence for the tissue on which the virus was first isolated and grown (Enders et al., 1962; Meyer et al., 1962; Katz, 1965; Schumacher et al., 1972b). The reverse is also generally true, virulent strains failing to grow in chick embryo cells, but there is evidence that all laboratory isolates are not alike, for the non-attenuated Philadelphia 26 strain produces plaques in Grivet monkey cells, whereas unattenuated Edmonston virus does not (Buynak et al., 1962).

Measles virus strains which have been isolated from patients with subacute sclerosing panencephalitis have a variable host range (p. 146); for example, two strains with different growth rates had each a wider host range in tissue cultures of monkey, human, dog and chick cells than had Edmonston virulent virus, but, apart from the higher yield of virus, they were similar in host range to an Edmonston vaccine strain (Horta-Barbosa et al., 1970). The opposite, a restricted host range, was found by Katz et al., (1970b) testing only two SSPE strain. It is not certain whether these differences are originally intrinsic, whether they represent selection by different methods of isolation or whether they represent different degrees of adaptation towards a common end-type of laboratory virus. That is, the stability of host range as a marker has scarcely been studied.

CYTOPATHIC EFFECT

Two types of alteration to cellular morphology were noted from the beginning of the cultivation of measles virus, besides the formation of intra-cytoplasmic inclusion bodies which itself came as something of a surprise to those who were growing the virus (Enders, 1956; Enders and Peebles, 1954). The first and most typical change in shape consisted in the appearance of multinucleate cells in tissue culture not unlike the giant cells known to Warthin (1931) and Finkeldey (1931)* and found also in monkeys infected with measles (Gordon and Knighton, 1941; Sergiev et al., 1960, 1966; Yamanouchi et al., 1970, 1973; and see Arakawa, 1964). The second alteration, found originally as a residual effect of syncytial formation or at the margin of the inoculated cell sheet, consisted of retraction and spindle cell or strand formation. Thought at first to represent a variant form of measles (Ruckle-Enders, 1962), this effect proved to be nutritionally dependent in some cultural conditions, for added glutamine could suppress the formation of giant cells and lead to spindle or strand formation (Reissig et al., 1956; Frankel and West, 1958). However, it became clear from the work of Oddo that concentrated inocula regularly produce strand formation and dilute inocula giant-cell formation, the effects being reversible on changing the dilution of the inoculum (Oddo et al., 1961, 1967). This sequence of events has been produced by other workers, e.g. Norrby et al., (1970).

Physiological conditions do affect the expression of cytopathogenicity, as is best seen in plaque formation, but low multiplicity of infection

* Earlier descriptions than these are referred to by Roberts and Bain (1958).

(Black, 1959b; Toyoshima *et al.*, 1959b; Nakai *et al.*, 1969), stasis of growth of the monolayer (Toyoshima *et al.*, 1959b) and variation in species of cell (Black *et al.*, 1956; Kohno *et al.*, 1968; Toyoshima *et al.*, 1959a) all influence the extent and type of cytopathic effect seen. It is not known whether the action of different measles variants is represented in these various cytological changes.

Differences have been noted in the type of CPE, syncytial or non-syncytial, and its rate of onset in two strains from subacute sclerosing panencephalitis (Hamilton *et al.*, 1973), but there is no evidence that the characters were stabilized at the time of testing. Gould *et al.*, (1976) have recently derived two types of measles virus from SSPE strains in which syncytial formation, plaque size and haemadsorption are correlated (see Fig. 3.3).

Minor differences between intra-cellular effects of measles virus have not proved reliable as an indication of strain difference. Cell type (Black *et al.*, 1956; Toyoshima *et al.*, 1959a; Rapp *et al.*, 1960; Nakai *et al.*, 1969), degree of adaptation (Matumoto *et al.*, 1961; Musser and Slater, 1962) and multiplicity of infection (Nakai *et al.*, 1969), as well as duration of the infection, all influence the proportions of inclusion bodies that are seen to be distributed between the cytoplasm and the nucleus of the cell. As a general rule cytoplasmic inclusion bodies appear sooner than intra-nuclear inclusions and the appearance and location of virus antigen, as revealed by specific fluorescent antibody, follows the same sequence as, and usually coincides with, the inclusion bodies in the cytoplasm (Mutai, 1959; Toyoshima *et al.*, 1960b).

Inclusion body-forming strains of measles virus and non-forming strains are therefore not known to exist, although some cell lines fail to show inclusion bodies when infected by measles virus (Black *et al.*, 1959).

One cytopathic change seemed rather alarming when it was first discovered. This was the occurrence of chromosome breaks in human lymphocytes taken and cultured during measles or after live measles vaccine had been given (Nichols *et al.*, 1962; Nichols, 1963). Other authors either do not agree with this finding (Tanzer *et al.*, 1963) or point out that it happens with other virus diseases or vaccines, as in measles, to an extent which is not significantly different from normal (Harnden, 1964). However, chromosomal changes do occur in tissue culture cells infected with measles virus and the change need not be related to the virulence of the virus strain (Avakova and Rapoport, 1971; Radyisch and Zakharchenko, 1972). None of the reports suggests that variants of measles virus exist with respect to the ability to produce breakages in chromosomes.

PATHOGENICITY AND VIRULENCE IN ANIMALS

These characters are even more complex than the production of cytological changes or plaque formation and, except for attenuation of vaccine strains towards man and monkey, they have not been much used as genetic markers. Lack of a suitable small animal model of measles contributes to this lack of interest and neurotropism in rodents is one of the few pathogenic characteristics to receive attention. It is dealt with in detail in Chapter 8. Here it is sufficient to say (1) that measles strains may be adapted to growth in brain, particularly in the mouse or the hamster and that the previous cultural history may influence the process of adaptation; for example, passage in FL, continuous human amnion cells, favoured adaptation to neurotropism in mice (Matumoto et al., 1964), (2) different strains, including commercial strains of measles virus, have different powers of growing and producing damage, after intracerebral inoculation, but before being specially adapted to nervous tissue (Mirchamsy et al., 1972; Katow et al., 1973; Shishido et al., 1973), (3) adapted virus may be difficult to recover from brain by means of non-neural tissue culture, but is infectious upon intra-cerebral inoculation (Burnstein and Byington, 1968; Schumacher and Albrecht, 1970; Griffin et al., 1974), (4) successful infection by neurotropic strains freshly recovered from brain may be non-yielding (O'Brien et al., 1972), (5) measles virus recovered from cases of subacute sclerosing panencephalitis is somewhat more neurotropic than wild virus and in vitro infections may be virus-yielding or non-yielding (Lehrich et al., 1970; Cernescu et al., 1972; Draganescu et al., 1972), (6) claims to have isolated measles virus from man by intracerebral inoculation of mice are few (Arakawa, 1948, 1949; Greenham et al., 1974).

One proflavine-induced mutant of measles virus was found by Haspel and Rapp (1975) to induce hydrocephalus regularly in hamsters. This pathogenic effect had been noticed as a sequel of acute measles encephalitis in hamsters and mice, when it followed or was accompanied by acute ependymal destruction.

PLAQUE FORMATION

This limited spread of cytopathic effect gives a more accurate indication of difference between virus strains than uncontrolled cell destruction; both size and form of plaques vary from strain to strain and from one tissue to another. Again, as a mark of attenuation, the presence or absence of plaques has proven reliable and useful. Edmonston-derived vaccine virus gives small plaques on Grivet kidney and pinhole plaques on chick-embryo cells. The virulent virus gives large plaques on human amnion tissue culture, but none on the other two sorts of monolayer (Katz, 1965). Likewise, Schumacher et al., (1972b), testing wild virus and two vaccine strains, as well as two strains recovered from the brain of patients with subacute sclerosing encephalitis, recorded that plaque size on Vero cells was a significant factor in distinguishing one strain from another. Here the SSPE strains did not affect chick cells.

The most stable plaque character seems to be the large-plaque or large syncytial lesion as the character has bred true in several laboratories (Seligman and Rapp, 1959; Oddo et al., 1967; Norrby et al., 1970; O'Brien et al., 1972). Seligman and Rapp suggested that their glutamine-insensitive syncytium may represent a property of true wild-type virus. One technical difficulty, virus purification, has resulted from the variety of lesions seen in a monolayer heavily inoculated with measles virus. These variations range from large syncytium or plaque, to small plaque and minute red dot (Matumoto, 1966), or a similar dense focus (Gould, 1974; Fig. 3.1). In addition, non-visible foci of infection have been recognized by fluorescent antibody methods (Rapp et al., 1959) and McCarthy has referred to the hazard of purifying virus from plaques in the presence of invisible foci of infection (McCarthy, 1962). Nevertheless, some stable plaque variants may exist. Gould and Cosby (unpublished observations) feel confident that some small plaque variants breed true at limit dilution, giving large syncytia or plaques only at low dilutions of inocula and the content of these large plaques gives only small plaques at limit dilution. If the reasons for the appearance of large plaques at high multiplicities are genetic, they may be of great importance. The finding is quite different from the loss of giant cell formation which occurs on repeated passage of undiluted virus (p. 27).

HAEMAGGLUTINATION

The ability of laboratory strains of measles virus to agglutinate red blood cells of certain species of monkey provided an apparently stable all-or-none characteristic. Erythrocytes from many other mammalian species fail to react with measles virus in this way, for example, rat, guinea-pig, rabbit, sheep, fowl and human group O cells (Peries and Chany, 1960) to which Rosanoff (1961) added mouse, cat, horse and cow cells. The haemagglutinating component of the virus is usually the virion, but disrupted viruses yield haemagglutinating material of which a second form was first detected by Schluederberg and Nakamura (1967). This small particle haemagglutinin agglutinated monkey red blood cells only in the presence of a high concentration of salt, especially

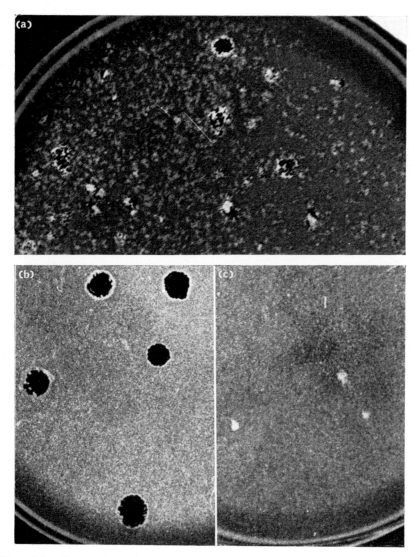

Fig. 3.1. Separation of a measles stock virus, Edmonston strain, into two plaque variants. (a) Portion of a monolayer of Vero cells showing large plaques and small dense foci. (b) Plaque-purified, large plaque strain at limit dilution in Vero cells; 5 large plaques, no dense foci. (c) Plaque purified, small focus-forming strain on Vero cells; 3 dense foci, no large plaques. (Gould, 1974).

ammonium sulphate, and could, unlike the standard virus agglutinin, be largely eluted by restoration of the saline to normal osmotic pressure. No whole virus counterpart of this haemagglutinin was known and it first occurred to Shirodaria, who was studying the infectious titre of a non-agglutinating strain of measles virus, that salt-dependence might also be a characteristic of the whole virus haemagglutinin. Several such strains were found in one laboratory and he and Gould have equated these to previously reported non-haemagglutinating, but syncytial-forming, virus, e.g. Enders-Ruckle strain 1677, and have suggested tentatively that wild measles virus may be in the salt-dependent state rather than possessing the standard haemagglutinin of the adapted laboratory isolate (Shirodaria et al., 1976). No studies to test this on fresh isolates have been undertaken, but salt-dependent variants have been isolated from two strains of subacute sclerosing panencephalitis virus. The property seems stable and promises to be a useful genetic marker (Gould et al., 1976). No other varieties of haemagglutinin have been described.

Haemadsorption

The presence of virus haemagglutinin at the cell surface is a natural part of the maturation process and haemadsorption may even be a more sensitive method of revealing virus antigen at that site (Fig. 3.2). Measles virus-specific cytotoxic antibody is an even more sensitive indicator of surface antigen when labelled with $^{125}I_2$ (Hayashi et al., 1973). The distribution of antigen varies in different situations. It is evenly distributed on the surface of individual infected cells and haemadsorption is found in the centre of measles-induced syncytia in tissue culture, often at a time when the amount of antigen present is insufficient to give good fluorescent-antibody staining. Kohn (1962), using KB cells, records haemadsorption on the syncytium as being peripheral; Gould using a plaque-purified salt-dependent strain found that salt-dependent haemadsorption covered the whole plaque and that a standard haemagglutinating virus gave central haemadsorption only in normal saline, but overall haemadsorption at higher molarity. This important observation implies that the standard HA strain is readily throwing salt-dependent variants or that the periphery of the plaque represents a different stage of development from the centre (Gould et al., 1976; Fig. 3.3). Other than this example, so far as we know, there are no stable variants showing haemadsorption patterns which differ from one another, but absence of haemadsorption has been used to characterize certain persistently infected, non-yielding cell lines, especially those carrying neurotropic measles virus strains (Payne et al., 1969; O'Brien et al., 1972).

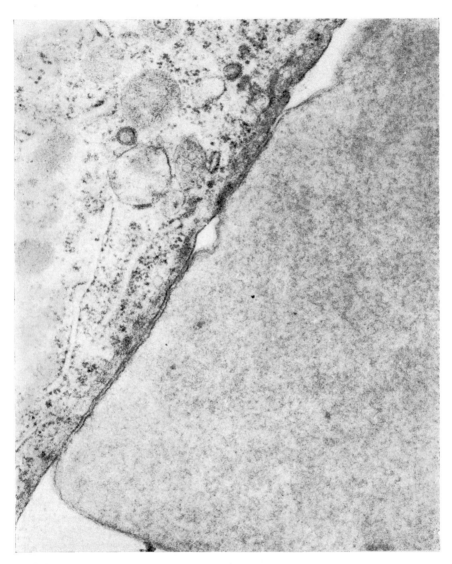

FIG. 3.2. Haemadsorption by a persistently infected HEp₂ cell in the absence of spikes. Magnification ×40,000.

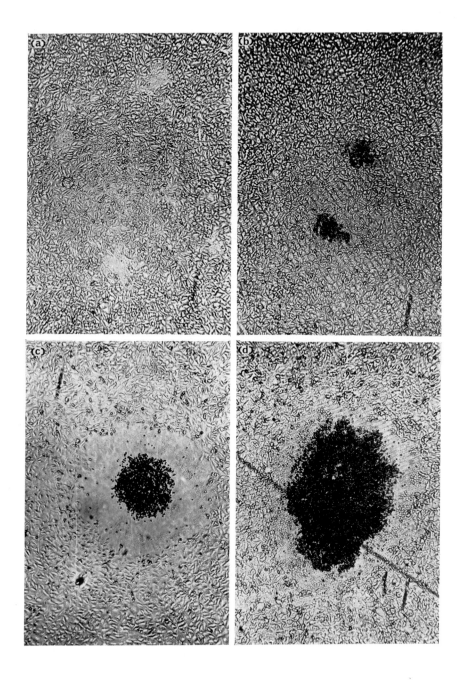

Haemolysis

This property of measles virus is relatively weak and dependent on adsorption of virus haemagglutinin to the red cell; it is therefore also specific for agglutinin-sensitive monkey cells. Some preparations vary in their powers of haemolysis relative to the agglutinating titre, and haemagglutinin and haemolysin have different degrees of sensitivity to destructive or inhibiting agents and are antigenically dissimilar (Norrby and Falksveden, 1964; Neurath and Norrby, 1965).

In spite of this, only one distinct variant from the normal haemolytic ability exists. A recent report establishes that mutants have been isolated which fail to synthesize haemolysin, as judged by an immunofluorescence test (Haspel *et al.*, 1975a,b).

TEMPERATURE-SENSITIVE MUTANTS

Strains of measles virus which are temperature sensitive have been recovered from persistently infected cultures (Haspel *et al.*, 1973) and a temperature-sensitive persistently infected line of HEp$_2$ cells was described by Gould (1974). It is not certain that these "naturally occurring" variants are stable, but induced temperature-sensitive mutants have been prepared (Haspel *et al.*, 1975a; Bergholz *et al.*, 1975) and should be of the greatest use in studying the replication of measles virus.

The demonstration of complementation groups in stable t-s mutants probably indicates that this method of analysing gene structure and function will be very useful with measles virus. It may be of some significance for the mitogenic effect of fluorouracil or to the mutability of particular cistrons that two sets of workers (Bergholz *et al.*, 1975 and Haspel *et al.*, 1975a) have each found three complementation groups and that mutants have been found which, as a result of defective maturation, cause the accumulation of measles virus antigens intra-cellularly, particularly capsid antigen. Analogy with SSPE strains was commented on.

Fig. 3.3.   Types of haemadsorption by measles virus-infected Vero cells. (a) Small plaque virus from LEC strain SSPE. Haemadsorption carried out with 1% monkey erythrocytes in isotonic buffered saline. Result negative, 5 plaques visible. (b) The same virus as (a). Haemadsorption carried out in 0·4 M ammonium sulphate in buffered saline. Result positive in two plaques. (c) Large plaque-forming viruses from LEC strain SSPE. Haemadsorption in isotonic saline. Result positive in centre of syncytium only. (d) The same virus as (c). Haemadsorption in 0·4 molar ammonium sulphate. Result positive up to edge of syncytium. (Gould et al., 1976).

Mutants from the same stocks of virus showed slightly different patho-genic effects on intra-cerebral inoculation, this being the first demon-stration that the postulated single step mutants of measles virus have different neuro-pathogenic potentiality (Haspel *et al.*, 1975b). However the first group of authors detected a degree of thermal sensitivity in their 9 mutants, whilst the others had selected 24 mutants which were heat-stable. It seems unlikely, apart from one mutant, that sensitivity of the virion to heat accounted for loss of pathogenicity of any of them, and Haspel *et al.* raise the question of defectiveness rather than tempera-ture-sensitivity at 37°C, which is the hamster's body temperature, as the real cause of decreased neuropathogenicity. The particular case of one mutant showing a selective form of neurotropism that resulted in the onset of acute hydrocephalus is referred to on p. 29.

Similar studies on relationships of different measles virus variants to the many different populations of leukocytes would be of great interest in view of the need to understand the effects of the virus on immune function (Chapters 12 and 13).

Temperature-sensitive variants have been selected by cultivation at low temperature. Makino and his colleagues (1970) state that they found these derivatives of Edmonston virus by using sheep kidney cells in culture at temperatures of 33°C, 29°C, 27°C and 25°C; Cernescu and his colleagues (1973) claimed that reduced temperature alone is insufficient to select low temperature variants but a change of host-cell is also needed, which in their experiments was provided by cultures of guinea-pig kidney. Their paper records also that the lower the temperature of growth, the greater the incidence of intra-nuclear inclusions. Whether this is secondary to a delayed growth cycle is not clear. It is also noteworthy that the cold variants gradually lost neuropathogenicity and formed smaller plaques on *Cercopithecus* R_6CA cells (Cernescu *et al.*, 1973).

SSPE STRAINS

During the recovery of SSPE measles virus strains in tissue culture it is sometimes possible to show a sequence of events by which a strongly suppressed cell-associated virus shows first specific intra-cellular anti-gen, then giant cells, then giant cells with haemadsorption, then extra-cellular haemagglutinin and the release of infectious virus. So far, there is no evidence that any of these stages corresponds to the existence of a stable defective mutant of measles virus. The important issue for patho-logy is whether (temperature-sensitive) variants of measles virus exist which can maintain a persistent infection at body-temperature without virus antigen appearing at the cell surface or not synthesizing it at all.

Gould claims to have seen this experimentally in his persistently-infected HEp_2 cells in which antigen synthesis is suppressed at 39.5°C and resumed after only two passes at 35°C, but details have not been published.

PHYSICAL PROPERTIES

No structural variants of the virus are known. Electron dense and translucent particles seen together in infected cultures do not represent different virus strains, and size is not a remarkably reliable property of any enveloped virus. Some physical characteristics of measles virus have been studied. Differences in density are not known to be strain-specific but do indicate infectious and non-infectious material (Schluederberg, 1962; Schluederberg and Roizman, 1962; Norrby, 1964).

Heat-resistance has not been found to vary much, but it is claimed that hamster-brain-adapted virus is more heat-sensitive than wild virus and that re-adaptation to *in vitro* growth is accompanied by increasing heat-resistance of the virus. The same may apply to resistance to freezing and thawing (Albrecht and Schumacher, 1972).

MEASLES, CANINE DISTEMPER AND RINDERPEST VIRUSES

In spite of the inadequacy of our knowledge of the genetic characteristics of measles virus, it cannot be denied that measles, canine distemper and rinderpest viruses are closely related and a little apart from paramyxoviruses. The bulk of the published evidence has been summarized (Imagawa, 1968) and similarities amongst the three viruses occur in host range, pathological effects, antigenic composition, mode of growth and substructure of the three virions. We shall refer briefly to each of these and to the history of the immunological findings.

Host range. Each virus seems to be limited naturally to one host or group of related hosts, but dog, ferret, and bovine kidney cells in culture have all been described as susceptible to primary inoculation with endemic strains of each of the three viruses. Measles SSPE virus has likewise been transmitted to ferrets, dogs and calves. Rinderpest virus has been adapted to rabbits (Yamanouchi *et al.*, 1974), but measles virus does not grow in rabbit kidney cells. Nicolle (1931) made a curious, but rather inadequately supported claim, that canine distemper virus could survive in man and Dalldorf and his colleagues (1938) made the much more substantial observation that Rhesus monkeys were susceptible to canine distemper virus grown in ferret spleen and inoculated by different routes. In addition to observing the striking intra-nuclear lesions, these workers actually recorded recovery of the virus by passage in ferrets from monkey brain 23 and 52 days after inoculation. There was no natural spread of the disease amongst monkeys.

Pathogenesis. Careful measurement of inclusion bodies in giant cell pneumonia led Pinkerton and his two colleagues to suggest that measles was the cause of the condition, but the subtitle of their paper—"A lesion common to Hecht's disease, distemper and measles"—indicates that pathology preceded immunology in pointing to a similarity between these two virus diseases (Pinkerton *et al.*, 1945). The same thought occurred to Adams *et al.* (1956), working with ferret lung. Inclusion bodies occur in circulating neutrophil polymorphs and not lymphocytes of dogs, (Cello *et al.*, 1959). Plowright and Ferris (1959) noted that rinderpest and measles both show intra-nuclear inclusion bodies at a late stage of the disease and pointed out that both viruses produce stellate cyto-pathic changes and spindle cell formation. Rinderpest virus (Tokuda *et al.*, 1963) and measles virus (Osunkoya *et al.*, 1974a) both grow and produce cytopathic effects in cultured leukocytes. The close similarity in intra-cellular changes produced by measles and distemper virus in tissue culture was described in infected ferret kidney cells (Shaver *et al.*, 1964) and extended to ultra-structural morphogenesis in Vero cells for all three viruses (Breese and De Boer, 1973); their studies add the in-formation that immuno-electron microscopy of infected tissue showed all three viruses cross-reacting serologically. Finally, the lymphotropism of rinderpest virus in rabbits was shown to be closely similar to that of measles virus in monkeys (Yamanouchi *et al.*, 1974).

The immunological relationship of the three viruses was first noted as a neutralization of chick-adapted distemper virus by human gamma-globulin (Imagawa *et al.*, 1954; Karzon, 1955) and the relationship to measles was traced antigenically by Adams and Imagawa (1957a,b) and epidemiologically by Carlström (1957). Some protection against measles was conferred on 165 individual patients by vaccination against distemper virus (Adams *et al.*, 1959). Numerous studies on the protective and cross-reactive relationships of the respective antibodies followed (Bech, 1960b; Gillespie and Karzon, 1960; Warren *et al.*, 1960; Imagawa *et al.*, 1960; Carlström (with review) 1962; Roberts, 1965).

In general terms, the neutralization titres discriminate sharply between the virus strains (Shishido *et al.*, 1967) and rinderpest and meas-les viruses induce protection against distemper better than distemper virus immunization against the others. There is also cross-protection between rinderpest and distemper (Goret *et al.*, 1957, 1959). Measles immunization does protect against rinderpest (Plowright, 1962). By far the most significant recent information is the fact that the nucleocapsids of measles, distemper and rinderpest viruses are antigenically closely allied or even identical (Orvell and Norrby, 1974). This implies that the relationship between the three viruses is as close as that between influenza A viruses of man and animals.

Physical properties and structure. Distemper and measles viruses were shown by Palm and Black (1961) to be heat-labile and ether-sensitive, to be inactivated by ultra-violet light and by gamma irradiation, each virus with a target size of 42 nm. Norrby and his colleagues give the ultrastructure of distemper virus by negative staining as an enveloped particle 115–230 nm in diameter, with club-shaped spikes 13 nm long and the inner component or nucleocapsid about 18 nm across (Norrby et al., 1964). The similar morphology and morphogenesis of all three viruses—measles, canine distemper and rinderpest—has been referred to (Breese and De Boer, 1973).

It is apparent that Imagawa's (1968) conclusion still holds good. Once cultured, measles, canine distemper and rinderpest viruses can be distinguished from one another only by host range and by serum neutralization. Indeed the relationship between the viruses is, if anything, closer than was apparent up to that time.

COMMENT

Smallpox, chickenpox, measles and German measles are distinguished from many other virus infections by the way in which one attack confers solid durable immunity. This implies some stability of the antigenic constitution of the virus and an absence of such naturally occurring variants or serotypes of the virus as are seen in influenza A and B or in the large families of enteroviruses. Since all four viruses came to be cultivated, this conclusion has been borne out, for serological variants of variola virus, varicella virus, rubella virus and measles virus are virtually non-existent.

This being so, variation should be sought in other characteristics and, even though these may be expected to be inconspicuous because of the uniformity of the clinical condition, minor differences in virulence, in host range or in plaque morphology, which have been found, may serve to distinguish various isolates of measles virus from one another, leaving the question of how they may be related to wild-type measles virus unanswered.

Isolation is bound to involve some degree of selection from wild-type virus. Thus, when two laboratory strains of measles virus such as Edmonston and Philadelphia 26 were found to exert different effects on Grivet monkey kidney cells, the difference is meaningless, so far as the existence of one or two kinds of endemic virus goes, unless the methods of isolation were as alike as possible. Only the cells used for isolation of measles virus are likely to affect selection, for two papers have shown that nutrient medium, if complete, and the type of animal serum in it, have very little effect on the growth or the virulence of measles virus (Horvath, 1973; Romano and Scarlata, 1973).

Cytopathic effects are probably best studied in plaques or in syncytia and here observed differences have been maintained constant by propagation at limiting dilution of infectivity. A real difficulty arises if variation of plaque type appears, when a seed virus is inoculated at a higher multiplicity of infection than that which yields a single type of plaque. The production of large plaques from a "small-plaque-forming" virus, when a low dilution of the "pure" virus was applied to the monolayer, followed by the production again of small plaques only, when the contents of the large plaque are plated out, is a case in point. Either the large plaque contains a small amount of a "large-plaque-forming virus" in an excess of "small-plaque-formers", so that only the latter appear on plating, or new phenotypic effects appear at high multiplicity of infection arising from genomes which cannot infect on their own. The difficulty of preparing a genetically pure stock strain seems to be greater than usual with myxoviruses and much more attention should be paid to maintaining all variants by passage at limiting infectivity and to testing their properties at intervals so as to determine the stability of each.

Nothing in the epidemiology of measles supports the belief that neurotropic variants exist as a cause of encephalitis, nor have pneumotropic or enterotropic variants of the virus been described. If such mutants could be obtained by induced mutagenesis, they would be useful in studying the pathogenesis of the disease which occasionally shows a predilection for pulmonary or intestinal symptoms.

The haemagglutinin of measles virus exists in at least two forms, acting in different osmotic conditions, and the haemolysin has been reported as absent from two induced mutants. These would all be worth studying in relation to host range penetration and uncoating of the virus in various species.

Temperature-sensitive mutants should be specially useful in studying the replication of measles virus and pathogenesis and persistent infection, which seems to favour temperature-sensitivity of the virus.

4

Structure and components of measles virus

INTERPRETATION—(1) *Measles virus resembles paramyxoviruses in structure.* (2) *Six structural polypeptides have been identified.* (3) *Two are glycoprotein containing glucosamine and are surface proteins (MW 69,000 and 53,000).* (4) *The nucleocapsid can be split by trypsin into unequal fragments (MW 38,000 and 24,000).* (5) *A water-insoluble protein (MW 35,000–40,000) may be a membrane protein.* (6) *No polymerase has been found in the virion.* (7) *It is not known whether surface spikes are uniform or mixed in composition and function.*

THE VIRION

Electron microscopy

Until 1969 information about the structure of measles virus was limited mainly to morphological studies (Waterson *et al.*, 1963; Norrby and Magnusson, 1965; Finch and Gibbs, 1968; Nakai *et al.*, 1969). The virus resembles paramyxoviruses in that the central core contains helices and there are radially distributed projections on the outer membrane. Measles virus has been classified as a member of the paramyxovirus group (Waterson, 1962) although its unique relationship with canine distemper virus and rinderpest virus and the absence of neuraminidase may justify an independent classification (Waterson and Almeida, 1966). Electron microscopic examination of measles virus has been carried out mostly on impure specimens, showing particles which are highly pleiomorphic with diameters ranging from 100–300 nm (Fig. 4.1). The virus consists of an outer membrane covered in spikes 9–15 nm in length surrounding the tightly coiled helical nucleocapsid which has a diameter of 16–17 nm.

Physical properties

The buoyant density of measles virus has been determined by equilibrium sedimentation on CsCl or potassium tartrate gradients. The reported values lie in the range 1·224–1·24 g/cm^3 (Norrby, 1964; Numazaki and Karzon, 1966; Hall and Martin, 1973; Phillips and Bussell, 1973). Similar densities have been found for SV5 (1·23 g/cm^3 in potassium tartrate; Klenk and Choppin, 1969) and NDV (1·236–1·25 g/cm^3 in CsCl; Stenback and Durard, 1963; Young and Ash, 1970). Recent studies by Rima in our laboratory have shown that the density of measles virus on metrizamide is 1·19 g/cm^3. The variations in the density reported from various laboratories may be explained by the type of virus growth which is taking place. Chiarini and Norrby (1970) found that the density of measles virus in CsCl gradients was greatly influenced by the multiplicity of infection and the passage history of the virus inocula. Oddo *et al.* (1961) introduced the terms UP virus (Undiluted Passage) and DP virus (Diluted Passage) and showed that the buoyant density of UP particles was 1·20–1·21 g/cm^3 whereas DP particles had a density of 1·23–1·25 g/cm^3. More recently rinderpest virus has been shown to have similar properties having a density of 1·21 g/cm^3 at a multiplicity of infection of 1, whereas at a multiplicity of 0·001 the density of the released particles was 1·23–1·25 g/cm^3 (Underwood and Brown, 1974). Johnston, Rima and Martin (unpublished results) have also found that measles virus purified on potassium tartrate gradients contained two components when the virus was passed undiluted, whereas only the heavier 1·23 g/cm^3 component was present when virus was grown after dilution of the inoculum to 10^{-3}. The reason for this difference in density is not yet

---

Fig. 4.1.   Measles virus nucleocapsid. (a) Rare circular form from cell lysate ($\times 125,000$) (b) Fragmented nucleocapsid extracted from infected cells with NP40 and banded in a CsCl gradient. The nucleocapsid measures 17–20 nm diameter, has an internal diameter of 7 nm and a pitch of 6·6 nm. (c) Nucleocapsid of twice the normal 1μm length from cell lysate ($\times 75,000$).

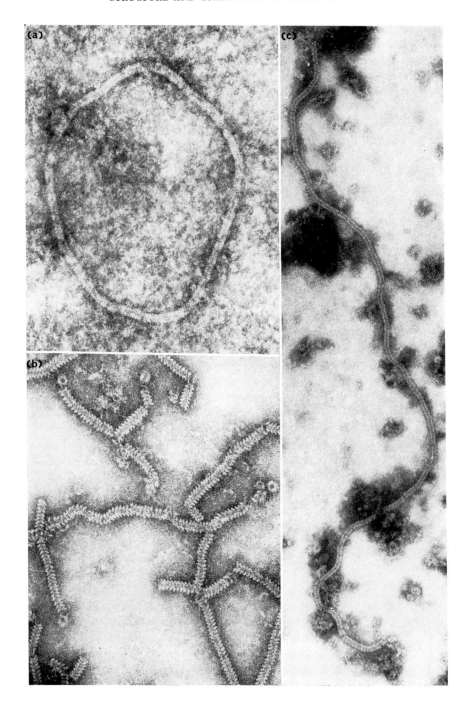

apparent as little precise chemical information is available on the particles produced by undiluted passage virus. However, it may be related to recent findings that nucleocapsids, only about one-tenth the length of the intact genome, can be incorporated into enveloped structures (Hall *et al.*, 1974), although the contribution of RNA to the density of enveloped particles is so small that this factor alone is unlikely to be a major determinant of these differences. It is also possible that the precise mechanism of maturation following UP or DP passages are different as observed with pseudo rabies virus by Ben-Porat *et al.* (1975).

The sedimentation coefficient of measles virus has not been estimated accurately (Hall *et al.*, 1974); however, the similarities between measles and the other paramyxoviruses make it probable that the sedimentation coefficient would be in the range 800–1000 S as reported for NDV and Sendai virus (Kingsbury *et al.*, 1970; Blair and Duesberg, 1970).

NUCLEOCAPSID

Nucleocapsids of measles virus contain the intact virus genome and are responsible for part of the complement-fixing (CF) activity of the virus particles. Nucleocapsids can be released from purified virus by treatment with 0·5% DOC and isolated on sucrose or CsCl gradients. Electron microscopy (Fig. 4·1) shows that the helix from infected cells has a characteristic herring-bone appearance with a diameter of 17 nm–20 nm and the hollow core, when viewed along the short axis of the structure, measures 7 nm.

Complex circular and branching forms of nucleocapsid exist of which the function in replication is not yet known (Thorne and Dermott, 1976 and unpublished observations). The buoyant density of nucleocapsids determined on CsCl gradients is 1·30–1·31 g/cm^3 (Hall and Martin, 1973; Norrby and Hammarskjold, 1972; Waters *et al.*, 1972; Kiley and Payne, 1975). Sedimentation coefficients of between 265–280 S have been reported (Hall and Martin, 1973; Norrby and Hammarskjold, 1972) but the methods used in extraction of the nucleocapsids have an important influence on the sedimentation behaviour. Electrophoresis on polyacrylamide gels of nucleocapsids, released by 0·5% DOC and purified by sedimentation on CsCl gradients, show that only a single species of protein was present having a molecular weight of about 60,000 (Hall and Martin, 1973). Reports by Underwood and Brown (1974) and Bussell *et al.*, (1974) have shown that the nucleocapsid proteins of canine distemper, rinderpest and respiratory syncytial virus are all close in size. Since the size of the nucleocapsid proteins of other paramyxoviruses,

NDV, SV5 and Sendai are also 60,000 daltons (Mountcastle *et al.*, 1970), the basic structure of all these nucleocapsids is probably very similar.

Ultra-violet spectrum analysis and direct chemical studies (Hall and Martin, 1973; Waters and Bussell, 1974) have shown that the nucleo-capsid of measles virus has an RNA content of 5%, the remainder being composed of a single species of protein of molecular weight approximately 60,000. A recent paper by Bussell *et al.*, (1974) showed that the method of isolation of nucleocapsids from cells had an important bearing on the morphology of the particles and size of the capsid protein. They isolated nucleocapsids from infected cells by use of proteolytic enzymes or simply by treatment of the cells with EDTA and freezing and thawing. When proteolytic enzymes were employed, the nucleocapsids were rigid and tightly-coiled helices whereas, if isolation was carried out in the absence of proteolytic enzymes, they had a more loosely-coiled and flexible appearance. These observations are consistent with the morphology seen by other workers who have isolated nucleocapsids from purified virions. Furthermore, they showed that the protein isolated from nucleocapsids prepared without proteolytic enzymes had a molecular weight of 60,000, as in virion material (Hall and Martin, 1973); however, when cells were harvested with trypsin, two proteins of molecular weights 39,000 and 24,000 were obtained. Similar observations have been made in our laboratory on measles virus nucleocapsid (Rima and Martin, unpublished observations).

The size of the nucleocapsids from infected cells has been studied by electron microscopy and sedimentation on sucrose gradients. Kiley *et al.* (1974) demonstrated that the nucleocapsids isolated from measles-infected cells were heterogeneous in length, although the ratio of RNA to protein remained the same. In order to obtain good resolution of the various size groups it was essential to treat the cytoplasmic extracts with both 5% DOC and Brij-58 prior to sedimentation on sucrose gradients. They demonstrated that the nature of the nucleocapsids present was related to the passage history of the virus inocula. Undiluted virus (UP) inocula produced a mixture of 200S and 110S nucleocapsids whereas diluted virus (DP) gave predominantly the 200S species. Furthermore, isolation of RNA from the purified nucleocapsid fractions yielded RNA directly related to the length of the nucleocapsid fragment and the 110S component contained RNA which sedimented at 18S and was only one-tenth that of the intact genome. These results are consistent with the observation of Carter *et al.* (1973b) that the smaller 20–18S RNA species predominates as infection proceeds. Also, they are in agreement with the observations that sub-genomic RNA can be encapsidated and eventually form incomplete or defective virion particles (Hall *et al.*, 1974; See p. 86).

VIRION RNA

The first report of the size of measles virion RNA was made by Schleuderberg (1971) who found the sedimentation coefficient was 52·2S relative to 50S RNA of SV5. Hall and Martin (1973) showed that the RNA sedimented at approximately 52S and was sensitive to ribonuclease. In all preparations of virus examined, a second component sedimenting at 4S was found, but this low molecular weight RNA was never found in purified nucleocapsids and was probably of cellular origin.

Reports from numerous laboratories have indicated a range of sedimentation values for RNA from paramyxoviruses in the range 50–57S. However, co-sedimentation experiments (Blair and Robinson, 1968; Compans and Choppin, 1968; Winston *et al.*, 1973) suggest that the virion RNAs have indistinguishable sedimentation values. Nonetheless, some variations in the sedimentation behaviour of RNA from measles virus and other closely related viruses have been noted by Norrby and Hammarskjold (1974). However, in contrast to the minor variations observed on sucrose gradient analysis, co-electrophoresis on polyacrylamide gels of measles, NDV, canine distemper and rinderpest virus RNAs shows no differences in the mobility. The variation in sedimentation values is probably due to the presence of large amounts of secondary structure in paramyxovirus RNA (Kolakofsky and Bruschi, 1973).

Probably the best estimate of the molecular weight of measles RNA is $6·4 \times 10^6$ (Fig. 4.2) which has been determined by electrophoresis on acrylamide–agarose gels relative to the RNA of bovine enterovirus (Todd and Martin, 1975) and ribosomal RNA (Petermann and Pavlovec, 1966). However, no information is available on the size of virion RNA after denaturation.

An unusual feature of paramyxovirus 50S RNA is its ability to self-anneal. Sendai virus RNA has been shown to anneal to variable extents (16% to 60%) (Portner and Kingsbury, 1970; Robinson, 1970). This implies that the 50S complementary strand of the genome can be encapsidated and can be incorporated into virions (Kingsbury, 1974). If this is the case with measles and other similar viruses, the comparison of base compositions of different viruses may not be of great value since the molar ratios will be an average of the two different strands present. Data on the base composition of RNA from purified virions is not available and a number of authors have encountered difficulties in obtaining sufficiently high activity in ^{32}P–labelled RNA preparations

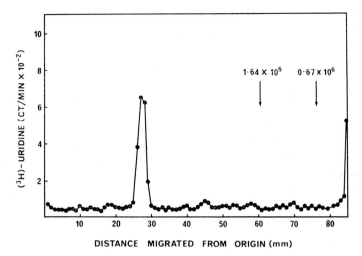

FIG. 4.2   Electrophoresis of measles virus RNA on 2·0% acrylamide-0·4% agarose gels. ●—●, ^{3}H-uridine, ct/min. The arrows indicate the extinction peaks of BHK ribosomal RNA expressed as molecular weight (Hall and Martin, 1973).

to permit analysis. The isolation of nucleocapsids from infected cells has, however, provided a richer source of RNA and analysis shows that the composition is high AU and relatively low GC ratios (Bussell et al., 1974). Table 9.12 compares the base composition of the nucleic acids of measles and related viruses (See p. 153 and p. 154 for comment).

In summary, the RNA of measles virus shows considerable similarities to that of other well-characterized paramyxoviruses. The infectious genome is an intact single strand with a molecular weight of about $6·4 \times 10^6$. The RNA is encapsidated by approximately 2,000 protein sub-units. It is very probable that both complementary strands are present in preparations of virus and, as will be discussed later (p. 89), sub-genomic RNA molecules may also be encapsidated. It is also very probable that large virus particles contain multiple nucleocapsids (Hosaka et al., 1966; Dahlberg and Simon, 1969; Granoff, 1962; Choppin and Compans, 1970) some of which may be complementary strands of virion RNA (Kolakofsky et al., 1974) or mixtures of sub-genomic nucleocapsids.

STRUCTURAL POLYPEPTIDES

Until 1973 no reports on the polypeptide composition of measles virus were available. Recently, two laboratories have been considering this problem independently and reports by Hall and Martin (1973) (Fig. 4.3) and subsequently by Waters and Bussell (1973) showed that the polypeptide composition was similar to other paramyxoviruses. Both studies involved the separation of ^{35}S-methionine or ^{14}C-amino acid labelled protein isolated from purified viruses by electrophoresis on SDS-polyacrylamide gels. These initial reports indicated the presence of at least six polypeptides in the molecular weight range 80,000–35,000

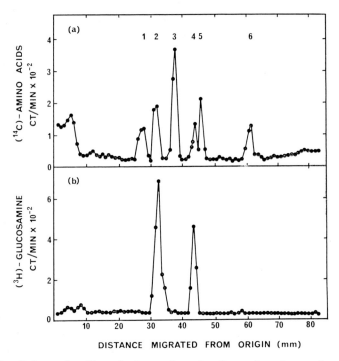

FIG. 4.3. Polyacrylamide gel electrophoresis of measles virus polypeptides. ●—● (a) ^{14}C-amino acids, ct/min: (b) ^{3}H-glucosamine, ct/min. (Hall and Martin, 1973).

(Pl–P6). There was also evidence of high molecular weight material which did not enter 7·5% gels, especially when virus was not degraded exhaustively with mercaptoethanol. Bussell et al., (1974) have also reported that a high molecular weight polypeptide of approximately 185,000 daltons is present and is the major component when virus particles are dissociated in the absence of a reducing agent. Their results suggest that the 185,000 component is a major virus protein containing two polypeptides which are linked by disulphide bonds as reduction results in the appearance of smaller polypeptides which co-migrate with P1 and P2.

Other reports on the molecular weight of measles virus proteins compare them with canine distemper and rinderpest viruses (Underwood and Brown, 1974). The information available is summarized in Table 4.1 and shows that considerable variation exists in the reported values

TABLE 4.1

Molecular weights ($\times 10^3$) of polypeptides of measles, canine distemper and rinderpest virus

Measles[a]	Measles[b]	Measles[c]	Canine Distemper[b]	Rinderpest[c]
Large	Large	110	Large	98
76	79	79	79	79
69	71	73	70	75
60	61	57	61	66
53	53	50	54	48
51				
46	46	41	46	43
	36	34	37	37
				29
				20

[a] Hall and Martin (1973); [b] Waters and Bussell (1973); [c] Underwood and Brown (1974). (Data assembled from Underwood and Brown 1974.)

especially for the faster moving polypeptides. However, these variations have probably little significance in themselves and are most likely a reflection on the experimental conditions used in the various laboratories. A much more thorough and comprehensive study is required before any statement concerning the differences in size of proteins from various strains can be made.

Nonetheless, some real confusion has arisen over the distribution of the glycoproteins present in measles virus. The initial report (Hall and Martin, 1973) showed that P2 and P4 were labelled with ^{3}H-glucosamine. More recently, Bussell et al. (1974) have reported that measles virus grown in the presence of ^{3}H-fucose as a carbohydrate precursor, resulted in only P1 being radioactive. Using the same labelling conditions, they also found that canine distemper virus contained two glycoproteins, namely P1 and P5. It is not possible at present to resolve this problem, but the information indicates that results obtained using single-carbohydrate precursors ought to be interpreted with caution. We have investigated the composition of amounts of concentrated measles virus which are sufficiently large to permit staining of carbohydrate components and although P2 and P4 are the predominant bands, other lesser staining bands are present. However, with large-scale preparations it is often difficult to assess the degree of purity and it is possible that the minor glycoprotein components are of cellular origin. In summary, at least two glycoproteins of approximately 70,000 and 53,000 daltons are present in measles virus and the presence of other minor glycoprotein components remains to be established.

Furthermore, our recent investigations have shown (Rima and Martin, unpublished observations) that following large scale preparation of non-radioactively labelled measles virus there are multiple protein species present which were not seen in our earlier studies on radioactive protein. The problem would appear to be related to the necessity to concentrate virus from large volumes of tissue culture fluid, when the amount of proteases present is greatly increased compared to the amount present during the processing of relatively small volumes of radioactively labelled virus. It has already been mentioned that nucleocapsid protein is susceptible to trypsin and, as we shall see, the surface proteins are readily released by mild treatment with bromelain. The apparent discrepancy in the reported molecular weights of the proteins of measles, CDV and rinderpest viruses is probably a reflection of the methods used rather than an indication of any fundamental difference in the various strains studied. This is a research area where more rigorous comparative studies are required. It should also be remembered that under certain growth conditions, isolated virus may contain host cell proteins that are either incorporated into virions or are absorbed by virus specific sites especially during concentration procedures where the relative amount of host protein to virus material is high. This problem is of particular relevance to viruses such as measles which does not grow to an exceptionally high titre (approx 10^6–10^7 p.f.u./ml), see note added in proof p. 54.

Nucleocapsid protein

As mentioned above the nucleocapsid protein (P3) has an approximate molecular weight of 60,000. However, it is susceptible to trypsin treatment and can be degraded to two components of molecular weight 38,000 and 24,000 respectively (Bussell *et al.*, 1974).

Membrane protein

After treatment of paramyxoviruses (NDV or SV5) with Triton X-100 in the presence of high salt (1M KCL) the virus glycoproteins, a non-glycosylated membrane protein and the envelope lipids are solubilized;

however, subsequent lowering of the salt concentration by dialysis results in the selective precipitation of the membrane protein. We have found that a protein in measles virus is also insoluble at low ionic strength (Hall and Matirn, 1974b) and so can be selectively precipitated from the soluble envelope components on removal of KCl by dialysis (Fig. 4.4).

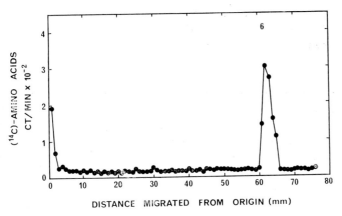

FIG. 4.4. Polyacrylamide gel electrophoresis of measles virus membrane protein. ●—● ^{14}C-amino acids, ct/min. (Hall and Martin, 1974b).

Surface proteins

The chemical structure of viruses can often be studied by the stepwise degradation of the particles by use of proteolytic enzymes or mild detergents. In the case of enveloped viruses this has been achieved by treatment with proteolytic enzymes such as bromelain or by detergents such as Tween 20, Triton X-100, or NP140. The rationale of the experiments to be described on measles virus has closely followed the protocol set out by Maeno et al. (1970) and Chen et al. (1971) and others in their studies on the structure of Sendai virus, SV5 or NDV. For the sake of clarity, the following description will be confined to the envelope of measles virus, although similar results have already been established in other paramyxoviruses.

Hall and Martin (1974a) showed that when purified measles virus was incubated with bromelain at 37°C for 4 h and the mixture centrifuged at 65,000 g for 1 h at 4°C the pellets obtained were devoid of the outer projections. Also, the supernatants obtained in such an experiment contained most of the HA activity, whereas HL and cell-fusion activities were both destroyed. The polypeptide composition of the spikeless components showed that four main proteins were present none of which was labelled with tritiated glucosamine (Fig. 4.5). Hence, it would appear that the glycoproteins are involved in the spike structures and are associated with HA activity. The sizes of the bromelain-released components were estimated by fractionation on sucrose gradients and by gel-filtration on Sephadex G200. Sedimentation in the presence of catalase showed that the virus material was approximately 6S, while gel-filtration showed

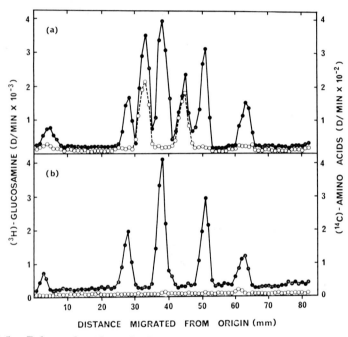

Fɪɢ. 4.5. Polyacrylamide gel electrophoresis of the polypeptides of (a) purified measles virus; (b) bromelain-treated virus. ●—●, ^{14}C-amino acids, d/min; ○—○, ^{3}H-glucosamine, d/min. (Hall and Martin, 1974a).

that the glycoprotein had a molecular weight of about 120,000. These results suggest that the two glycoproteins present in the envelope are associated strongly even after release from the virus. Electron microscopy of tween–ether-released components shows them as rosettes (Fig. 4.6).

COMMENT

In conclusion, we can now say that there are at least six structural proteins associated with measles virus as detected by radioactive labelling techniques, although other minor components may be present which would not be detected by the present analytical methods used. A very thorough investigation involving perhaps the use of iodination techniques, which greatly increases the sensitivity of detection, may allow minor components to be resolved. Clearly, if the nucleocapsid of an intact

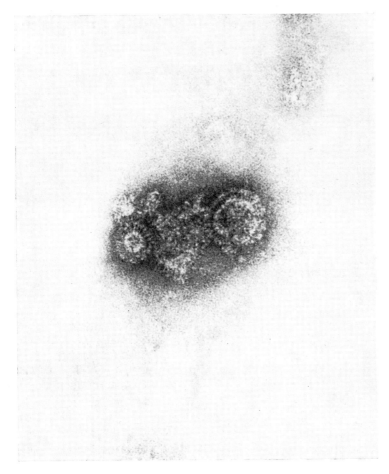

FIG. 4.6   Rosettes of 12·5 nm spikes on membrane fragments which result from Tween 80-ether treatment of measles virus.

virion contains about 2,000 sub-units, then it would be highly improbable that other proteins present only in small numbers (such as 50) would be detected by present methods of analysis.

The structural features of measles virus derived by the biochemical dissection of purified virions are summarized in Fig. 4.7.

The nucleocapsid sub-unit has a molecular weight of 60,000 and is sensitive to cleavage by proteolytic enzymes to fragments of approximately 38,000 and 24,000 daltons respectively.

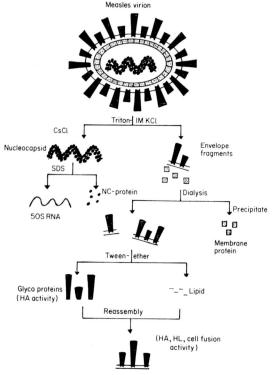

FIG. 4.7. Schematic presentation of the structural components and the biochemical dissection of measles virus. HA = haemagglutinin; HL = haemolysin.

Note added in proof. Rima *et al.* (1977) have recently reassessed the number and nature of the proteins in measles and canine distemper viruses by use of pulse-labelling with ^{35}S-methionine for short periods throughout the growth cycle. This technique allows a distinction to be made between host and newly synthesized viral proteins. The only viral glycoprotein detected has a mol. wt. of 69,000 whereas the previously reported glyco-protein (53,000) appears to be a host component. In addition to the four major virus-induced polypeptides, namely: 69,000 (glyco-protein), 65,000 (P2), 58,000 (nucleocapsid protein), 37,000 (M–protein), we find consistently a small polypeptide of 15,000 daltons. Furthermore, the 43,000 component, which is a consistent feature of all reports (see Table 4.1) is also a host protein which has very hydrophobic properties.

Our current studies indicate that a number of specific host proteins are intimately associated with the envelope of measles and CDV particles.

5 | Non-vegetative functions of measles virus, haemagglutination, haemolysis and cell-fusion

INTERPRETATION—(1) *Measles virus has extracellular effects similar to those of paramyxoviruses in respect of haemagglutination, haemolysis and cell-fusion but not neuraminidase activity.* (2) *The functions have not been shown to control entry or release of virus to and from cells except by implication and in cell fusion.* (3) *They are useful in studying morphogenesis of the virus envelope.*

Certain *in vitro* effects of measles virus are not connected with the vegetative phase of virus growth. They are functions of the virion and its component parts which are not recognizable *in vivo* and which have to be examined by artifice. These effects are haemagglutination and haemolysis of monkey erythrocytes and fusion of cultured cells by extracellular virus. It is likely that cell-fusion from without, which can also be effected by non-infectious virus, is a result of the same process which causes giant cell formation *in vivo* and in cell cultures, but the syncytia in infected cells are not necessarily produced by whole virus particles.

Haemagglutination and haemolysis were first described adequately by Periés and Chany (1960, 1961, 1962) in a series of papers giving the main properties of the virus and of the red cells which are involved in the two reactions and to which very little of equal significance has since been added. Haemadsorption by measles virus-infected cells was demon-

strated earlier by Mastyukova and Khait (1960) who may also have
seen specific haemagglutination. Cell fusion has been well described by
Cascardo and Karzon (1965) and has hardly been investigated since
then.

HAEMAGGLUTINATION

Measles virus agglutinates the red blood cells of certain primates,
excluding man, but including several kinds of old-world monkeys.
Although readily available species from several other mammalian families
have been tested, their erythrocytes are insusceptible to agglutination
by measles virus (Periés and Chany, 1960; De Meio and Gower, 1961;
Rosanoff, 1961; Kohn, 1962). This is a point of some interest, considering
the fact that cells of several of these species have proved susceptible
to infection in the laboratory and a few of the animals, e.g. dog, ferret,
mouse and hamster, can also be infected by measles virus either before
or after a course of adaptation of particular strains (chapter 2). Contrarily,
we have in the laboratory several strains of measles virus which do not
ordinarily agglutinate monkey red blood cells but which readily infect
monkey kidney cells in culture (Shirodaria et al., 1976). There are different opinions about the sensitivity of erythrocytes from different species
of monkey. Rosanoff, using $\frac{1}{2}\%$ cell suspensions and 0·5 ml volumes of
reactant, found that rhesus, cynomolgous, patas, green monkey and
baboon red blood cells were equally sensitive in the haemagglutination
test (Rosanoff, 1961). Cells from baboons (cynocephalus) and patas
(erythrocebus) monkeys were preferred originally (Periés and Chany,
1960, 1962) but Funahashi and Kitawaki (1963) found that, out of several
individual cynomolgous monkeys tested, there was a range of titre given
by the same virus stock from 1:64 to 1:1024 and, between species, red
cells of the green monkey gave the highest average titre. Enders-Ruckle
(1965) gives Cercopithecus red blood cells as the most sensitive of four
species tested, when compared with cynomolgous, patas and rhesus
species.

It was clear from the beginning that monkey red blood cells could
absorb infectious virus as well as haemagglutinin from a measles virus
suspension, but all haemagglutinin was not virus, for the two properties
could be readily separated by filter membranes of pore-size 25, 47 or
56 mμ (Periés and Chany, 1962). Other indirect evidence of two physical
states of the haemagglutinating property came from studies of resistance
to physical agencies and the effects of centrifugation. For example,
centrifugation at 26,360 g for 1 h sedimented infectivity and a portion
of the haemagglutinin and left a less dense fraction of the haem-

agglutinin in the top quarter of the tube (Periés and Chany, 1961); Norrby (1962a) observed a sharp change in the slope of heat-inactivation curves which suggested that two different molecular populations of haemagglutinin must exist. In consequence, ratios of infectivity to haemagglutinating power have not been determined exactly, varying from $10^{3 \cdot 0}$ to $10^{4 \cdot 0}$ TCD_{50} per haemagglutinating unit (Norrby, 1962a) to at least $10^{5 \cdot 0}$ to $10^{6 \cdot 0}$ TCD_{50} per haemagglutinating unit (Periés and Chany, 1960). The accuracy of the two values is not helped by the fact that different methods have been used to concentrate virus sufficiently to give measurable amounts of haemagglutinin. Centrifugation methods may lose haemagglutinin, forced dialysis may concentrate small particle haemagglutinin equally with whole virus and both methods cause loss of infectivity.

Adsorption of virus to red blood cells is a relatively slow process, taking at least 30 min to reach equilibrium (Norrby, 1962a), and all workers agree that the highest agglutination titres are obtained at 37°C (Periés and Chany, 1960; De Meio and Gower, 1961; Funahashi and Kitawaki, 1963). Once absorbed, measles virus agglutinin is not eluted by change of pH within the range pH 4·0 to pH 10·0 nor by changing the temperature and molarity of the suspension although ions, especially divalent ions, are needed for adsorption of virus to the red cell surface (Norrby, 1962a). Cells treated with formalin were not agglutinable by measles virus (Periés and Chany, 1961, 1962). Receptors on the red cell were destroyed readily by trypsin but not by potassium periodate (Norrby, 1962b) nor by receptor-destroying enzyme or by treatment of red cells with myxoviruses (Periés and Chany, 1962; Norrby, 1962b). The haemagglutinin itself is trypsin-sensitive and periodate-sensitive (Norrby, 1962b; Periés and Chany, 1962; Funahashi and Kitawaki, 1963), but is relatively stable at 4°C for three months (Periés and Chany, 1962) or 22 days (Funahashi and Kitawaki, 1963). Recently elution has been effected by suspension in a solution of arginine at a molarity of one (Lebon *et al.*, 1975). Haemagglutination can be reversed by adding specific immune serum (Norrby, 1962a).

Only a few reports exist on the biochemical characteristics of the HA component of measles virus (Hall and Martin, 1974a,b; 1975). Treatment of the virus with 0·5% v/v Tween 20 (or NP 40) in 0·02 M bicarbonate buffer (pH 10·0) at 37°C for 1 h offers a convenient way of solubilizing the envelope components. The lipid components can be separated by centrifugation on CsCl gradients where the HA activity was obtained as a single peak at a buoyant density of 1·26 gm/cm^3 and was independent of the presence of lipid. However, both glycoproteins were present in these preparations and attempts to identify which one is directly related to HA activity was made by treatment of the virus with Triton X-100 in 1 M KCL under conditions where the protein lipid complex was not completely disrupted. After removal of the membrane protein (p. 50) and subsequent centrifugation on sucrose gradients, two envelope fractions could be partially separated. One of these contained the larger glycoprotein, although the slower sedimenting fraction always contained a portion of the larger as well as most of the smaller component. Table 5.1 shows the recovery of HA activity at various stages of the Triton treatment. Solubilization of the virus results in a two-fold increase in HA activity and this is presumably due to conformational changes in the envelope structure. The recoveries of HA activity in the final sucrose gradient indicate that the activity is associated with the larger glyco-protein; however, no HL activity could be detected in either fraction. This evidence suggests that the two glycoproteins on the surface of measles virus are responsible for the biological activities, HA, HL and cell fusion. The larger glycoprotein contains the active site for HA, but it is not yet established whether the HL activity resides solely in the smaller glycoproteins. In contrast to the HA activity, the presence of lipid is important for haemolytic activity.

TABLE 5.1

Haemagglutinating activities of fractions of measles virus obtained by treatment with Triton X-100 followed by centrifugation on three consecutive sucrose gradients

Sample	HAU	Presence of Glycoproteins	
		Large	Small
Purified virus	512	+	+
Triton-solubilized envelopes	1024	+	+
Fast sedimenting component	64	+	−
Slow sedimenting component	1024	+	+

(Data from Hall and Martin, 1974b.)

A haemagglutinin is known to exist which agglutinates only at a higher salt concentration than that of normal saline. It was first described as a non-integrated substance recovered from suspensions of virus by centrifugation and acting best in hypertonic, buffered ammonium sul-phate, molarity 0·8 M, pH 7·3 (Schluederberg and Nakamura, 1967).

It is now clear that intact virus can possess the salt-dependent property as a heritable character recognized in recently isolated viruses and in plaque purified preparations (Shirodaria *et al.*, 1976; Gould *et al.*, 1976). This raises the question as to whether the salt-dependent or the normal haemagglutinin is contained in wild measles virus, for non-haem-agglutinating strains, which could be or could include salt-dependent strains, have been found by other authors (Waterson *et al.*, 1963). Disruption of such virus did not yield conventional HA. Other varieties of haemagglutinin may exist, for Tischer (1967) noted that measles virus from which agglutinin to intact monkey red blood cells had been fully removed by absorption with untreated erythrocytes still contained some agglutinin which reacted with RDE-treated cells. In addition, certain preparations of salt-dependent haemagglutinating virus require high salt molarity to agglutinate, but unlike Schluederberg's agglutinin, these preparations do not elute from red blood cells when osmotic normality is restored (Gharpure, unpublished observations).

The biochemical basis of SDA is unknown, but in our laboratories a comparison of the proteins associated with HA and SDA strains is being made. A clear understanding of the possible strain differences may help to resolve some of the contradictory reports concerning the sizes of the various glycoproteins from different viruses such as measles, canine distemper and rinderpest, especially since the latter two are not haem-agglutinating viruses.

Finally, these observations fail to establish any clearcut relationship between infectivity of measles virus for tissue culture and the ability to adsorb to and agglutinate red blood cells. This is unlike myxovirus in which destruction of receptors reduces infectivity (Stone, 1948) and it does suggest that haemagglutination may be an artefact of laboratory adaptation. If that were so, naturally occurring or wild measles virus would resemble salt-dependent haemagglutinating strains more closely than standard agglutinating preparations of the virus.

HAEMOLYSIS

Partial lysis of erythrocytes appeared originally in haemagglutination tests carried out at 37°C with red blood cells of patas monkeys or of baboons (cynocephalus) and, after 4 hours at 37°C, visible haemolysis equalled haemagglutination in titre. We have confirmed this using rhesus monkey erythrocytes, 1% or 2% suspension, in barbiturate buffer, and a laboratory-adapted strain of Edmonston virus (unpublished observations). De Meio (1962) chose a 1% suspension of rhesus red blood cells as the most sensitive indicator of haemolysis, demonstrated a pH optimum at pH 7·5 and noted that the haemolytic activity was not always proportional to the haemagglutinating titre: haemolytic activity could also be inactivated at a temperature of 50°C, at which temperature haemagglutinin was not affected. Norrby and Falksveden (1964) resorted to forced dialysis in order to obtain sufficient virus to examine dose–response curves of the haemolytic action; the plot had an upward curve. They found that the percentage haemolysis of a fixed concentration of erythrocytes rose with temperature to 40°C, the pH optimum was pH 8·0 and the haemolytic activity of the preparation was neutralized by measles-specific antiserum. Freezing and thawing increased the haemolytic activity of their preparation, a fact reminiscent of haemolysis by NDV and mumps virus (Chu and Morgan, 1950). Haemolysis was somewhat inhibited by the presence of calcium ions and the property was abolished by treatment of the virus preparation with ether.

The same upward-curving dose–response for haemolysis by measles virus that was seen by Norrby and Falksveden (1964) was noted also by Saburi and Matumoto (1966), who observed that a given dose of virus always haemolysed the same proportion of cells in a suspension, within certain limits of cell-concentration. Taking one haemolytic unit as the amount of virus which gives 25% haemolysis of 2% cells after 2 h incubation at 37°C, they constructed neutralization curves with measles antisera. The curves were regularly sigmoid and placed parallel to each other at distances proportional to antibody concentration. The test seemed to be slightly less sensitive in their hands than the anti-haemagglutinin test and both were less sensitive than the neutralization test.

Neurath and Norrby (1965) were able to find reactions which selectively inhibited haemolytic activity of measles virus, as compared with haemagglutination, particularly photo-inactivation by methylene blue and they found some retardation by diisopropylfluorophosphate, but the exact nature of the haemolytic effect was not elucidated.

Incubation of measles virus-infected cells with glucose analogues during the growth cycle of the virus is said to interfere with glycosylation of the proteins of the envelope as with other enveloped viruses (Kaluza et al., 1972; Hodes et al., 1975*). In recent work on measles virus-infected cells Shirodaria has found that small doses of glucosamine hydrochloride, 2 mM–8mM concentration, abolish the haemolytic activity of the virus and, incidentally, syncytial formation by infected cells, whilst the production of haemagglutinin at the cell surface continues (Fig. 5.1). There is a marked drop in the yield of infectious virus (Shirodaria, unpublished observations).

Thus a more direct link seems to exist between infectivity, cell fusion

* This paper contains no data on measles virus, only a statement.

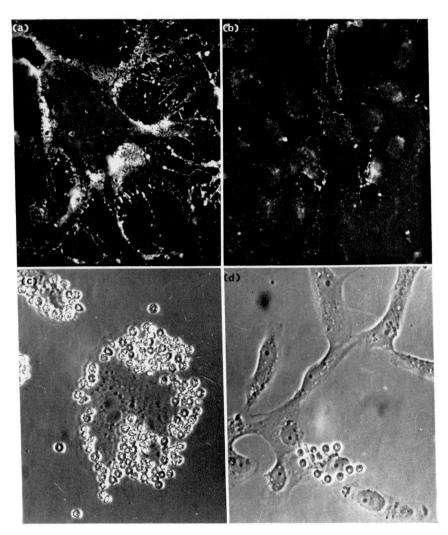

Fig. 5.1. The inhibitory effect of glucosamine hydrochloride on the replication and function of measles virus. HEp$_2$ cells infected with measles virus at m = 3 and incubated at 35°C with and without glucosamine hydrochloride. (a) and (c) without glucosamine hydrochloride show syncytial formation much surface staining of virus antigen by immuno-fluorescence of unfixed tissue and strong peripheral haemadsorption. (b) and (d) with 6 mM glucosamine hydrochloride show inhibition of syncytial formation, reduced virus antigen and much less intense haemadsorption (Shirodaria et al., unpublished).

and the ability to cause haemolysis, than there seems to be between infectivity and haemagglutination, although it must be understood that without haemagglutination, haemolytic potential cannot be detected.

CELL FUSION FROM WITHOUT THE CELL

Giant cell-formation was the first *in vitro* cytopathic effect of measles to be recorded (Enders and Peebles, 1954). The power of inactivated measles virus to cause giant cell formation was shown by Toyoshima and his colleagues (1960a) using a continuous line of human amnion, FL, cells.

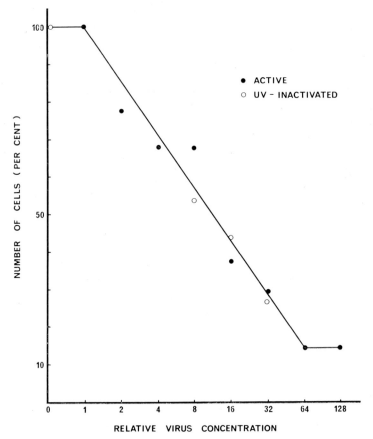

FIG. 5.2.   FL cells incubated with increasing concentrations of measles virus, or the same concentration of virus irradiated with UV light to ten times more than the dose needed for complete inactivation. The falling number of cells, measured 6 h after infection, indicates cell fusion. (Slightly adapted with permission, from Toyoshima *et al.*, 1960a.)

UV-inactivation was applied at a dose ten times greater than that needed to inactivate the infectivity of the virus preparation, but there was no loss of the ability to fuse FL cells (Fig. 5.2). Virus that had been heat-inactivated at 56°C for 30 min lost the fusion property which, with UV-inactivated and with uninactivated virus, reached its maximum at 3 h.

Measles-specific antibody neutralized the fusion effect when added to virus-cell mixtures up to 90 min after adsorption of virus to cells had begun, but was ineffective thereafter.

Cascardo and Karzon (1965) re-examined the fusion effect, using Edmonston virus at a concentration of $10^{8 \cdot 3}$–$10^{9 \cdot 5}$ TCD_{50}/ml and a different strain of human amnion cells (AV3), which were tested at the 3rd or 4th day after plating. This technique showed fading of inter-cellular boundaries at 30 min after adding virus and complete fusion of cells at 1 h. Live virus or UV-inactivated virus (4,200 ergs/sec/cm²) had like effects. The nuclear membrane was also disrupted after 9 h of ex-posure to virus. The fusion effect was more resistant to heat-inactivation than was haemolysis by the same preparation of virus, but serum neutralization of infectivity and fusion decreased together, though with different inactivation slopes. Different types of cells had different sensitivities to fusion factor. Little else seems to have been done to determine whether haemolytic and fusion factors are identical nor to discover the mechanism of susceptibility to fusion factor, but Poste et al. (1972) were satisfied that the different sensitivities of species of cell to haemolysis were not related to the respective ratios of cholesterol or lipid in each type of cell membrane.

Some knowledge of the role that the various components of the en-velope play in haemolytic and cell fusion activities was obtained by reassembly experiments (Hall and Martin, 1974a,b). When the lipid and glycoprotein components separated by isopycnic sedimentation on CsCl gradients were mixed and dialysed against PBS for 4 days at 4°C, haemolytic and cell-fusion activities were restored (Table 5.2). These restored activities were of viral origin as they were inhibited by anti-measles serum. Artificial complexes of the virus glycoprotein and specific lipids, such as phosphatidylethanolamine, phosphatidylserine, phos-phocholine, sphingomyelin and cholesterol, have also been studied by Hall and Martin (1974b). The ability of individual virus phospholipids to form active haemolysin varied considerably. Cholesterol did not form active HL but could enhance the activity produced by reassembly of glycoproteins and phosphotidylethanolamine. This suggests that choles-terol interacts with the other phospholipids rather than the glycoprotein. The observation that a variety of lipids can be used to regenerate

haemolytic activity also suggests that the role of the lipid is essentially
passive, and is probably required to hold the glycoprotein in a biologically
active conformation. Similar conclusions have been made for SV5 by
Hosaka and Shimizu (1972).

TABLE 5.2

Reassembly of membrane components

Origin of Lipid	Quantities of components dialysed for 4 days			Properties of reassembled components		
	Glycoprotein		Lipid-P			
	HAU	μg	μg	HAU	HL (%)	Cell Fusion
Virus	128	31	10·2	128	59	+
Virus	128	34	15·4	128	64	+
Virus	256	48	15·0	256	68	+
Virus	256	51	17·8	256	71	+
Virus	512	74	20·0	512	76	+
Phosphatidylethanolamine	256	49	50·0	256	48	+
Phosphatidylethanolamine	256	46	40·0	256	47	+
Phosphatidylethanolamine	256	45	20·0	256	33	NT
Phosphatidylethanolamine	256	49	8·0	256	14	±
Phosphatidylethanolamine	256	44	2·0	128	0	—

NT = Not tested. The membrane components of measles virus were separated into the glycoprotein and lipid fractions by centrifuging in CsCl. Mixtures of the glycoprotein and virus lipid
or phosphatidylethanolamine were dialysed for 4 days and the HA, HL and the cell fusion
assayed. (Data from Hall and Martin, 1974a.)

Analysis by acrylamide gel electrophoresis of the proteins required
for reassembly of haemolytic activity indicated that both the large and
small glycoproteins were necessary. Whether or not the HL activity
resides solely in the smaller sub-unit is not known.

Preliminary evidence is available that suggests that haemolytic activity
is the result of proteolytic action (Hall and Martin, 1974b). Measles virus
HL activity can be suppressed by phenyl methyl sulphonyl fluoride, a
protease inhibitor. In contrast, haemolysis induced by lysolecithin is not
inhibited by concentrations of the inhibitor which can suppress measles
virus haemolysis. It would seem, therefore, that the virus proteins are
principally responsible for haemolysis and not a lipid component such as
lysolecithin as previously proposed by Rebel *et al.* (1962) and Barbanti-
Brodano *et al.* (1971).

COMMENT

It is likely that the larger glycoprotein is an active HA unit, but HL activity can only be expressed when both glycoproteins are complexed in a particular manner with virus lipid. It is possible that a better understanding of the role of the surface components can be obtained by considering the intact envelope rather than attempting to relate the individual components to convenient properties such as HL and HA activities. The real function of the envelope is presumably to attach to, and penetrate, the cell membrane. The present evidence suggests that this process involves a receptor protein (HA sub-unit) which binds to the appropriate cell receptor sites. In an intact envelope (or envelope fragment) the second glycoprotein will be held in the correct juxtaposition by the lipid to ensure that proteolytic activity can initiate the process of membrane fusion. The hypothesis that the smaller glycoprotein may have the active site of proteolytic activity can be readily tested now that the large glycoprotein can be obtained in a pure form. On the other hand, it could be that the active site for cell lysis is on the larger glycoprotein and the smaller one acts as an allosteric activator. These intriguing problems are of a general nature, relating to the paramyxovirus field, and await an early solution.

6

The growth cycle of measles virus

INTERPRETATION—(1) *The growth cycle of measles virus resembles that of paramyxoviruses.* (2) *It differs from them in the important respect that some function of the cell nucleus is essential for complete virus replication.* (3) *Reasonably accurate growth curves of measles virus can be constructed allowing for difficulties imposed by certain properties of the virus.* (4) *The latent period may be as short as 14 hours and the eclipse period 8 hours.* (5) *Virion RNA (52S) is not usually found free in infected cells.* (6) *Smaller RNA molecules (18S, 27S) may be synthesized faster than large RNA species.* (7) *Early virus proteins have not been characterized.* (8) *Much more information is needed about details of RNA synthesis and virus assembly.* (9) *Defective interfering components are readily produced by passage of undiluted virus suspensions.*

GENERAL FEATURES

Accurate construction of a growth-curve for measles virus has been handicapped by three prominent characteristics of the virus: (1) It usually produces a poor yield of infectious virus compared to many other RNA viruses (Table. 4.1). (2) There is a distinct and steady loss of infectivity at incubator temperature (Black, 1959b) once the virus is released from the cell. (3) A moderate amount of autointerference accompanies attempts to increase the amount of virus grown or to achieve synchronous infection by raising the multiplicity of infection (Toyoshima et al., 1960b; Matumoto et al., 1965; Nakai et al., 1969). A fourth complication also arises from a function of the virus and affects the accuracy of the assay rather than the total amount of virus recovered, and that is its ability to cause cell fusion and so reduce the number of infectious centres when applied externally to cell cultures at a high ratio of virus particles per cell (Toyoshima et al., 1960a).

Before considering steps of the growth cycle separately, some general points can be made. As techniques have improved sufficiently to ensure that one-step growth conditions are attained, it seems that the eclipse phase can be made as short as 10 h. Indeed the paper by Shishido et al. (1967), see Table 4.1, quotes an eclipse period of 6 to 12 or 8 to 10 h on FL cells, given in unpublished data of S. Kohno. The shortest eclipse periods are found at high multiplicities of infection which are more than large enough to ensure that all cells are infected. When lower multiplicities of infection are used, but still sufficient to ensure that all cells in the culture are synthesizing measles virus, as judged by the fluorescent antibody method, longer eclipse periods are found (Matumoto et al., 1965). This suggests that the mean eclipse time, like that of influenza virus, can be reduced by multiple infection of all cells and that the shortest eclipse phase may well be 10 h or less. Several authors have reported that the yield of infectious virus declines at the higher multiplicity of infection (Underwood, 1959; Matumoto et al., 1965; Nakai et al., 1969). It is well known that continued passage of undiluted virus also results in low yields of infectious virus, a state of affairs analogous to autointerference with other RNA viruses (Oddo et al., 1961, 1967).

Not many growth cycles have been sampled at short enough intervals to measure accurately the time interval between first appearance of cell-associated virus and the first virus released, but Shirodaria's results, given in Fig. 6.1b, show a delay of about 6 h and an eclipse period as short as 8 h. Oddo's experiments show the shortest time lapse between the end of the eclipse phase and the end of the latent period (Oddo and Sinatra, 1961). An early paper of Toyoshima's has a difference of 2 days (Toyoshima *et al.*, 1959b), but the multiplicity of infection used in the experiments is not stated. Very low input of measles virus may not cause a visible cytopathic effect for as long as three weeks after infection, but sub-culture of the infected cells resulted in immediate virus replication with the appearance of cell destruction (Toyoshima *et al.*, 1959b).

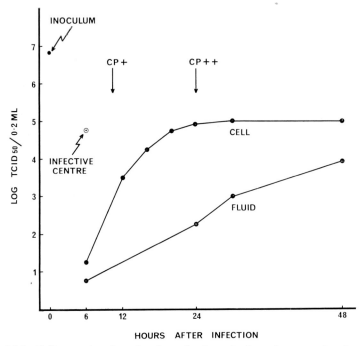

FIG. 6.1A. Cell associated and released virus during the growth of measles virus (Toyoshima strain) in FL cells. Inoculum $10^{6.8}$ TCID$_{50}$ per 5×10^5 cells. Infectious centres reduced by cell fusion. (Slightly adapted, with permission, from Toyoshima *et al.*, 1960b).

Total yields of infectious virus tend to be low as compared with other RNA viruses, the highest recorded being $10^{7.5}$ to $10^{8.0}$ TCD$_{50}$ per ml as mentioned by Mutai (1959) from monkey-adapted strains on human cell lines and by Schluederberg (1962) using Edmonston virus on HEp$_2$ cells. Highest yields are found at lower temperatures of incubation than 37°C, e.g. 33·5°C (Underwood, 1959).

THE ASSAY

Although plaque production by measles virus was successfully demonstrated within a few years of the virus being isolated (Hsiung *et al.*, 1958),

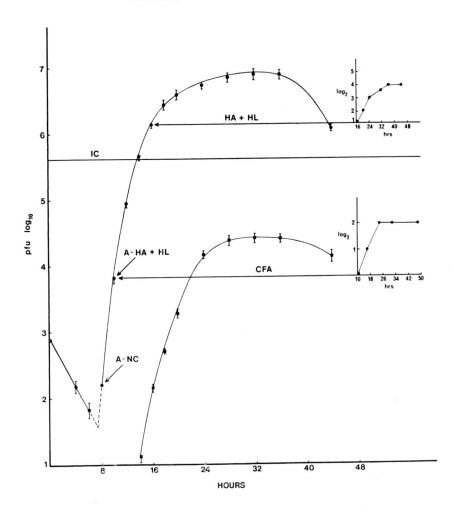

Fig. 6.1B. Growth of Edmonston-derived measles virus strain 243 in Vero cells infected at a multiplicity of 3 and assayed every 2 h for different virus functions. p.f.u. = plaque forming unit; HA = haemagglutinin; HL = haemolysin; CFA = complement fixing antigen; NC = nucleocapsid; IC = infectious centre; A = antigen by fluorescent antibody. Arrows indicate the time at which HA, HL and NC first became visible by immunofluorescence. The shortest release period indicated by the graph is 6 h. The yield of virus is about 10·3 p.f.u. per infectious centre. A different strain, P9, shows identical sequences, but a yield of > 90 p.f.u. per infectious centre. Lower curve: released virus; upper curve: cell-associated plus released virus (Shirodaria, unpublished observations).

TABLE 6.1

Data on the growth of measles virus in cell-culture

Reference	1. Cell type 2. Virus strain	Multiplicity "m"	Adsorption period	Virus titres Cell associated (per ml except stated)	Released	Eclipse Period (shortest times)	Latent Period	Remarks
Black, 1959b.	1. HEp$_2$ 2. Edmonston	1·5	2 h 37°C			18 h	26 h	
Underwood, 1959.	1. HeLa 2. Edmonston	10	4 h	N.D.	$10^{5.5}$ TCD$_{50}$	N.D.	N.D	Daily sample
Toyoshima et al., 1960b.	1. FL 2. Toyoshima	3·0 to 0·03	2 h 35°C	$10^{5.0}$ TCD$_{50}$	$10^{2.0}$ TCD$_{50}$	12 h	24 h	Titres with low "m" higher yield.
Oddo and Sinatra, 1961.	1. HeLa 2. Edmonston	3·6 to 20·3	1 h 22°C	$10^{5.5}$ TCD$_{50}$	10^{5} TCD$_{50}$	12–16 h	16 h	Multiplicity measured. Proper plateau in 24 h.
Kohn and Yassky, 1962.	1. KB 2. Edmonston	0·025	(room temperature 60 mn: in suspension)	$10^{5.0}$ TCD$_{50}$	$10^{3.1}$ TCD$_{50}$	16 h	24 h	Low multiplicity.

Reference	Cell / Virus	Multiplicity	Adsorption	Titre	Peak	Titre	Peak	Remarks
Matumoto et al., 1965.	1. FL 2. Sugiyama	0·03 4·4 and 38·0	Room temp. 15 mn plus 1 h 36°C	$10^{6.8}$ p.f.u. per bottle	—	$10^{5.8}$ p.f.u. per bottle	10 h	18 h	Cloned virus to get rid of "red spot" plaques. One step growth.
Shishido et al., 1967	1. Vero 2. TyOSA (Toyoshima)	1·4	1 h 37°C	$10^{5.8}$ TCD$_{50}$	$10^{4.2}$ TCD$_{50}$	18 h	24 h	
Nakai et al., 1969.	1. BS-C-1 2. Edmonston	0·01 1·0 and 10·0	2 h 37°C	$10^{5.7}$ p.f.u.	$10^{5.3}$ p.f.u.	—	24 h	Higher titres at low "m". Sampling times not given.
Norrby, 1972	1. Vero 2. Edmonston	3·0	4 h 37°C (rocked)	$10^{5.5}-10^{7.0}$ TCD$_{50}$	Lower than C.A.V.	20 h	N.D.	Growth curve incidental.

quantal assay persisted for some years more (examples occur in Table 6.1). This may partly explain why information about the various steps of the replication cycle is still scarce and incomplete.

Plaque assay

In the plaque assay, proportionality with virus concentration has been shown (De Maeyer, 1960; Arita and Matumoto, 1968).

Plating efficiency on different cell lines has not been studied much, but Karaki (1965b) gives descending plaque numbers on HEp_2, KB, FL and HeLa cells for the strain Sugiyama, adapted to the FL cells. He noted that 4-day-old cultures, which still have dividing cells, gave higher plaque numbers.

The development time for measles virus plaques seems to vary from 6 days to 10 days, after which the number does not increase, nor does the size (De Maeyer, 1960; Karaki, 1965b). The size and shape of the plaques can vary within single virus preparations on the same monolayers and can vary also from one cell type to another as well as from strain to strain of measles virus. Small opaque plaques and minute red dots have been recognized as foci of infection from the early days (McCarthy, 1962; Matumoto et al., 1965) and re-recognized by Gould (1974). These less well-developed forms frequently outnumber distinct plaques and have caused all these observers to remark on the hazards of virus purification by the plaque method and the need to eliminate them as potential interfering agents before attempting precise assays of infectivity (See Fig. 3.1).

Virulent and attenuated viruses have been described as making circular or elongated plaques respectively on kidney cell cultures from the Grivet monkey (Buynak et al., 1962) and several authors have purified strains of viruses which make large syncytia or large plaques as a predominant characteristic (Oddo et al., 1961, 1967; Matumoto et al., 1965). However, stable strains of viruses which make only small or opaque or red foci on the same cell line in which they appeared alongside other more distinct plaques have not been described. There is a need to determine whether the origin of the incomplete plaques is genetic, physiological or a consequence of interference in multiply infected cells.

Although fluorescent antibody has been used widely as a means of detecting infected cells, it has not been generally adopted as a means of assay in spite of an excellent and early demonstration of its speed, accuracy and economy, although the authors did not think that at the time (Rapp et al., 1959). It should be particularly applicable to the study of non-plaque-forming infectious units of measles virus, if these exist, and also to assays of defective variants which produce abortive infections.

Infectious centres

Very few assays include the proper measurement of infectious centres, but such centres are interesting in the study of persistent "non-yielding" infection by measles virus (p. 99) when the number of cells capable of forming infectious centres may, paradoxically, be much greater than the number expected from the practical absence of infectious virus in the culture fluids (Knight *et al.*, 1972; Bather *et al.*, 1973). They can also be used to detect infected brain cells when the released virus is infectious by intra-cerebral inoculation, but not on tissue culture (Burnstein and Byington, 1968).

ADSORPTION

Most workers seem to have realized that firm adsorption of measles virus to monolayers is a rather slow process which was first measured by Underwood (1959) and found to take 4 h to complete on HeLa cell monolayers, being only 60% complete at the end of 2 h. De Maeyer (1960) reckoned that 2 h was sufficient for adsorption on HeLa cells, WS or human amnion cells, and apparently on chick embryo monolayers, although complete data were not given in his short report. Many publications refer to a period of 1 or 2 h adsorption as convenient and giving reproducible results (Black, 1959b; Nakai and Imagawa, 1969; Nakai *et al.*, 1969; Toyoshima *et al.*, 1960a,b), but Oddo and Sinatra (1961), using HeLa cells, and Kohn and Yassky (1962), who used KB cells, gave an exposure time of 1 h, apparently at room temperature, and Norrby (1972) returned to a 4 h period of adsorption on Vero cells and rocked the inoculum on the monolayer. In our own laboratory 4 h of adsorption gives a very reproducible proportion of infected cells from any one seed virus, as assessed by the fluorescent antibody method (Haire, unpublished data).

As with most viruses 0·2 ml is a suitable volume of inoculum for monolayers on conventionally sized plates. The proportion of adsorbed virus fell as the volume increased above 0·2 ml (Arita and Matumoto, 1968). The same authors have described how replacement of buffered saline with isotonic glucose as a suspending agent stops adsorption of the viruses, whilst mere depletion of the magnesium and calcium content of buffered saline reduces the plaque numbers significantly below the normal efficiency of plating. It has been shown that the number of measles virus plaques obtainable from a constant inoculum can be raised by about half when a hypertonic inoculum (430m Osm) is used as suspending agent (Hövel, 1971).

PENETRATION

Firmness of adsorption and penetration, as measured by increased resistance to the subsequent application of specific antibody to the monolayer, are both greatly temperature-dependent so that adsorption at room temperature may well benefit from longer adsorption times. In mouse L cells a very rapid adsorption and penetration time seems to have been detected by using an acid glycine buffer to terminate infection, but the ensuing uncoating of viruses was inordinately slow (Kohno et al., 1968). Palacios (1965), in a cytological study of the Pittsburgh strain of measles virus, could trace input virus at the cell surface 2 h after infection, but not at 4 h after infection. It is our experience that adsorbed virus at the cell membrane can be traced for 8 or 10 h, if the inoculum is concentrated, even when virus suspension has been carefully filtered to remove fluorescent staining debris and only the spotty or patchy distribution of the fluorescence prevents confusion with synthesis of new antigen which does not appear for some hours after that time (Shirodaria, unpublished data).

THE REPLICATIVE PROCESSES

General survey

In contrast to what is known of paramyxoviruses such as NDV, Sendai and SV5, very little information about the synthesis of measles virus RNA or protein is available at present. However, during the past few years sufficient data have accumulated to allow at least a superficial comparison of mechanisms of replication in measles-infected cells with the more thoroughly investigated members of the paramyxoviruses. The replication of paramyxoviruses has been described in several recent reviews (Blair and Duesberg, 1970; Kingsbury, 1972, 1974) and only the salient features will be mentioned here. On infection, the paramyxovirus envelope fuses with the cell membrane and the nucleocapsid structure is released into the cytoplasm. Recent evidence suggests that the intact nucleocapsid is the active structure for RNA transcription (Stone et al., 1972; Marx et al., 1974). Transcription is achieved by a virus RNA polymerase which can recognize specific bases in the 50S genome despite the fact that it is still associated with capsid proteins. The RNA polymerase is probably present in only a few copies per nucleocapsid structure and recent evidence by Marx et al. (1974) indicates that a polypeptide of molecular weight 75,000 may be associated with enzyme activity. Transcription results in the production of mono-cistronic messenger RNAs. Collins and Bratt (1973) have separated seven messenger RNAs species from NDV-infected cells by electrophoresis in acrylamide–diacrylate gels. The species were present in non-equimolar quantities and their molecular weights formed a series which was directly related to the molecular weights and amounts of the virus polypeptides.

The second role of the paramyxovirus genome is associated with its own replication which involves the production of an RNA strand complementary to the intact virus strand. This complementary strand, in turn, functions as the template for the synthesis of progeny RNA strands. The production of the complementary strand requires an RNA polymerase which ignores the transcriptional signals for the production of messenger RNA and reads through the entire genome.

There appears to be no free 50S RNA in Sendai virus-infected cells (Blair and Robinson, 1968; Robinson, 1971) and this suggests that 50S complementary RNA would also be encapsidated. 50S RNA from virus or cellular nucleocapsids was able to anneal to the same extent as 50S RNA from infected cells or virus particles (East and Kingsbury, 1971). This self-annealing would represent the formation of a duplex between the previously encapsidated 50S RNA and the 50S complementary strands. This may mean that nucleocapsids containing complementary 50S RNA may play an important role in virus RNA replication and would imply that the capsid proteins may have a function in replication. As infection proceeds, a balance between replication and transcription may be controlled by the availability of capsid protein (Kingsbury, 1974).

Replication of RNA in measles virus-infected cells

Parfanovich et al. (1971) drew attention to the importance of the passage history of measles virus for the nature of the RNA produced. The size of RNA induced in Vero cells following infection with DP (diluted passage) or UP (undiluted passage) variants (p. 42) was investigated by growing the virus in the presence of actinomycin D and ^{3}H-uridine. RNA synthesis reached a maximum at the fifth day after infection when there was a spread of cytopathic effect throughout the monolayer. Three types of RNA were found by sedimentation on sucrose gradients corresponding to 34–32S, 20–10S and a small population of 50S. When cells were infected with UP variants, the amount of label incorporated was markedly reduced and no significant cytopathic effect was noted. This report and earlier papers by Oddo et al. (1967) and Chiarini and Norrby (1970) served as a clear reminder of the need for caution in the use of virus harvests for one-step growth cycle experiments. In order to achieve a successful one-step growth cycle, it is necessary to passage the virus at high dilution for a number of times until a preparation of high titre is obtained and then to use this at a high multiplicity for only one experiment. This procedure avoids the difficulties of studying a system involving defective particles (see p. 86).

The most comprehensive study of viral RNA synthesis in measles-infected cells is provided by Carter et al. (1973b) who noted that sub-genomic RNA was found in preparations of measles virions made by undiluted passage virus but not after serial passage of diluted inoculum. In their experiments on the kinetics of RNA synthesis they ensured that the inoculum used contained virions with predominantly 52S RNA. Sedimentation analysis of measles virus specific RNA isolated from cells at different times after infection revealed several size classes. The main components sedimented at 20S, 27S and 35S. A minor 52S component, presumably virion RNA or its complementary strand, was sometimes clearly present, although not always found. By labelling at different times after infection Carter et al. (1973b) showed that the size classes of measles virus-specific RNA appeared in sequence. At 13 to 16 h post-infection there was no evidence of 20S material, but significant amounts of 35S and 27S RNA. From 19 h onwards there was an obvious increase in the relative amount of 20S until it became the predominant species present. This apparent time-dependent control of RNA synthesis has also been reported for NDV (Kingsbury, 1970). By the use of temperature-sensitive mutants, Carter et al. (1973b) showed that all three RNA species were completely inhibited by a shift to the non-permissive temperature. The evidence suggests that there is no disjunctive synthesis of the various size groups of RNA and the accumulation of the 20S RNA late in infection may be explained if the small RNA species are synthesized faster than larger segments.

Similar results have also been found by Hall (1974) in Vero cells infected with DP virus. Infected cells were labelled with [3H]-uridine in the presence of actinomycin D for 4 h periods between 14 and 18 h and 22 to 26 h post-infection and sucrose gradient analysis showed that the relative proportion of the smaller 18S and 27S RNA species increased significantly towards the end of the growth cycle. The majority of measles-specific RNA accumulating in infected cells was ribonuclease-sensitive indicating that it was single-stranded.

The bulk of the evidence on the synthesis of RNA in measles-infected cells is in general agreement with what is known about other paramyxoviruses. Namely, a number of monocistronic messenger RNA molecules are produced corresponding to the coding potential required for measles protein. A relatively small amount of virion-sized RNA is produced, however, and it would appear that the sub-genomic RNA may replicate more rapidly and accumulate late in infection.

The role of the nucleus in measles virus replication

In an early autoradiographic study of the sites of synthesis of viral and cellular RNA in measles-infected cells, Parfanovich *et al.* (1966) examined the uptake of [³H]-uridine in mouse embryo cells in the presence of actinomycin D. In cells from measles-infected cultures, incorporation of uridine within the first 4 h was inhibited in the nucleoli and, to a small degree, in the extra-nucleolar part of the nucleus. Five hours after infection the labelling in the extra nucleolar region increased dramatically together with apparently renewed synthesis of RNA in the nucleoli. Since we now know that the nucleolus plays an important role in the export of RNA to the cytoplasm (Harris, 1974) these results may imply that the nucleolus of measles-infected cells is also functioning.

Evidence for the involvement of nuclei in the growth of measles virus is also provided by the detection of antigens and nucleocapsids in nuclei of infected cells. Nakai *et al.* (1969) have noted that nucleocapsids, as a rule, appear first in the cytoplasm and then in the nucleus. There are also repeated findings of nucleocapsid masses in nuclei of brain cells in SSPE patients (ter Meulen *et al.*, 1972b). The relationship between nuclear inclusion bodies and virus replication has not been determined, but the intranuclear appearance of nucleocapsids is a property of measles, canine distemper and rinderpest viruses, which distinguishes them from the paramyxoviruses. The exact role of the nucleus in measles virus replication is not known, although it does not seem to be required for synthesis of virus antigen (Follett *et al.*, 1976), but there is other evidence which suggests that it is intimately involved. Recently Norrby (1972) studied the independent appearance of envelope and nucleocapsid components in infected Vero cells by immunofluorescence techniques. By use of a carbobenzoxy tripeptide which prevented the spread of virus from cell to cell (Norrby, 1971) he was able to study antigen production under conditions of a one-step growth cycle. It was observed that nucleocapsid antigen appeared in the cytoplasm 8 to 10 h post-infection. The intensity and amount of fluorescence increased until 16 to 20 h post-infection and at this time nuclear fluorescence was also observed. In contrast, envelope antigens were found exclusively in the cytoplasm and again were first detected at 8 to 10 h post-infection. The significance of these observations in regard to replication processes remains to be seen, but the accumulation of nucleocapsid antigen in the nucleus late in the growth cycle indicates that some compartmentalization of replication events may occur.

Carter *et al.* (1973a) have investigated the uptake of uridine into nuclear RNA in infected cells treated with actinomycin D. After prolonged labelling, viral RNA was found concentrated in the nuclear fraction although the site of accumulation may be perinuclear rather than intranuclear. Following short labelling periods, viral RNA with a size distribution similar to that expected for transcriptive intermediates was found in both the nucleoplasm and the particulate fractions. Wild *et al.* (1974) have shown similar-sized RNA fractions, some of which were RNAse-resistant, in nuclei of rinderpest-infected cells. Schluederberg and Chavanich (1974) have reported that, even when measles-infected cells are exposed to cordycepin (50 µg/ml) throughout the growth cycle, there is no inhibition of virus growth. It is of interest that Mahy *et al.* (1973) found that the replication of influenza virus showed a similar resistance to 30 µg/ml cordycepin, whereas NDV and Sendai virus were sensitive at concentrations as low as 12·5 µg/ml. Davidson and Martin (unpublished results) have also found that measles virus is unaffected by concentrations of cordycepin below 30 µg/ml.

Although there is still much confusion over the role of the nucleus in measles virus replication, the above information is sufficient to indicate that nuclear involvement with the replication of measles virus represents a distinct difference between it and other paramyxoviruses. If the data prove to be true, then we may be able to understand the growth of measles virus according to the following two-stage process. The replication process, which produces large 50S genome RNA, may be confined completely or partially to the nucleus. Both complementary strands may be encapsidated in the nucleus and move via the nucleolus to the cytoplasm where some strands are used as templates for the synthesis of messenger RNA. However, this transcriptive process may not be confined to the production of messenger RNA alone, but may also generate sub-genomic RNA of the same sense as the intact virion RNA. With the recent development of techniques to study virus replication in enucleated cells, it should be possible to examine this hypothesis in detail (Follett *et al.*, 1975, 1976).

Effect of actinomycin D on the replication of measles virus

Investigations on the replication of RNA viruses have been greatly facilitated by the use of the drug actinomycin D which can inhibit the synthesis of most cellular RNA, thus making it possible to detect virus-specific RNA by radioactive labelling procedures. Unfortunately, in many reports the specific function of actinomycin has been taken for granted and often insufficient information is presented to ensure that complete cellular inhibition has been achieved, or that other side-effects of the drug are not affecting the results (Martin and Brown, 1967a,b). It is important, at present, that we look critically at this information since it is becoming clear that the role of actinomycin is not always what is expected or desired by the research worker. For example, recently some picornaviruses have been found to be sensitive to actinomycin (Grado and Ohlbaun, 1973) and different cells may have different dose–response rates (Schluederberg, 1971), and, also, actinomycin does not completely inhibit the post-transcriptional addition of poly A residues to cellular RNA which becomes especially evident when [^{32}P] is used as a radioactive precursor (Martin, unpublished data).

The effects of actinomycin on measles virus growth are particularly confusing, with reports ranging from a clear enhancement of growth to suppression. Anderson and Atherton (1964) reported that actinomycin D at low concentrations (0.01–0.1 μg/ml) enhances the growth of the Edmonston strain of measles virus adapted to HEp$_2$ cells and grown in fibroblast monolayers, which gave an extremely poor yield of active virus when not treated with the drug. Matumoto et al. (1965) and Mirchamsy and Rapp (1969) found similar enhancement of the growth of various strains of measles in a variety of cells and interpreted their results by suggesting that the increase in measles titre was primarily the result of suppression of interferon production by inhibition of host-cell DNA-dependent RNA synthesis. These authors also noted the suppression of measles virus growth at concentrations about 1 μg/ml, but the effects noted were considered insignificant because of toxicity of the drug (Mirchamsy and Rapp, 1969; de Jong and Winkler, 1970). Vero cells have been shown to be relatively resistant to the toxic effects of actinomycin (Schluederberg, 1971) and concentrations up to 50 μg/ml have been used to obtain complete inhibition of cell RNA synthesis. Also, Vero cells have been reported to be deficient in interferon production (Desmyter et al., 1968) and therefore provide a useful culture system for investigating replication of measles virus.

Schluederberg et al. (1972) studied the effects of actinomycin D on the growth of measles Edmonston strain in Vero cells under conditions where host-cell RNA synthesis was completely inhibited and found that the yield of infectious measles virus was reduced by 99–99.7% by doses of actinomycin D of 1 μg/ml or greater. In fact, there appeared to be a dose-dependent response to both host-cell and viral RNA synthesis which were equally sensitive immediately after infection, although 24 h post-infection cellular RNA synthesis was more sensitive. The expression of cellular DNA did not seem to be a requirement for measles replication, as pretreatment of cells with actinomycin D had little or no depressive effect on the yield of virus. This is in keeping with earlier reports on measles virus growth in cells irradiated with UV light (Ackerman and Black, 1961; Rapp, 1964). The time of addition of actinomycin D after infection has an important bearing on the yield of virus. When added after the adsorption period, the yield of infective virus is reduced by 100-fold, but no reduction occurs in the production of HA and cell-fusion activities. On the other hand, if the drug is added 24 h after infection, it causes only a 3-fold decrease in the yield of infective virus.

Thus, the inhibition of infectivity noted during the first 24 h after infection cannot be explained by a general toxic effect, since other viral expressions—haemagglutination and cell-fusion activities—were not inhibited. Schluederberg et al. (1972) suggested that the selective inhibition of measles virus may be an effect on an early step prior to the completion of template for 50S genome and that actinomycin may act directly to inhibit the formation of replicative intermediates. However, it would seem that messenger RNA production is not affected since viral membrane proteins are made in the presence of the drug at both early and late times after infection. This proposed selective inhibition of measles RNA synthesis is not without precedent. East and Kingsbury (1971) found that synthesis of cell-associated mumps 50S RNA was suppressed by actinomycin D, but that synthesis of small complementary strands was unchanged or even enhanced although their conclusion that host-cell genome expression was essential for mumps maturation may not now be justified. Similar results have also been found in recent studies on canine distemper and measles virus (Martin, unpublished results) which suggest that considerable inhibition of virus RNA synthesis occurs at concentrations of actinomycin D necessary to inhibit Vero cell RNA synthesis. These findings with measles, canine distemper and mumps viruses indicate that resistance to actinomycin D is not a characteristic differentiating all paramyxoviruses from orthomyxoviruses as suggested by Blair and Duesberg (1970) and provides further justification for placing "measles-like" viruses in a separate sub-group from that represented by NDV and Sendai virus.

In summary, actinomycin D may have two contrasting effects on the development of measles virus, depending on the type of cell used and the concentration of the drug. Enhancement of virus growth may be achieved at very low concentrations when most of the actinomycin present will be complexed with cellular DNA and prevent the production of interferon. At higher concentrations, when there is probably excess of actinomycin D over the amount of drug required to block the DNA available, the drug may have a specific effect on the formation or replication of the replicative intermediate. It is conceivable that this involves the complexing of the drug with the double-stranded complexes formed during early replication steps. Why this does not appear to occur with other paramyxoviruses such as NDV or Sendai is not so clear, but it could be related to the locality in the cell of replication of the virus or to the presence of specific sequences which may favour the binding of the drug to a duplex RNA molecule. On the other hand, the preferential inhibition of influenza RNA by actinomycin is well documented and there are certain similarities with measles virus, although it is generally considered that influenza virus requires the function of host-cell DNA. Nevertheless, the differential inhibition of 52S measles RNA relative to the production of messenger RNA by actinomycin may be indicative of the cellular compartmentalization of replicative and transcriptive processes in measles virus growth as was mentioned on p. 77.

Synthesis of virus antigens and proteins

Antigen tracing. The appearance of virus antigens during the growth cycle has generally been followed by fluorescent antibody labelling techniques. Fluorescent, coccoid bodies, less than 1 µ in diameter first appeared at or just before 12 h after infection of HEp_2 cells, although the response of human amnion cells to the same strain of virus was found to be much slower. Subsequently specific measles antigens were detected in both the cytoplasm and nucleus, with a definite nucleolar association (Rapp et al., 1960). The exact specificity of the serum used for fluorescence was not reported, so it should be noted here that association of measles virus antigen with nucleoli is contrary to all subsequent experience except in one instance of double infection by tick-borne encephalitis virus and measles virus when measles virus antigen was seen to accumulate in nucleoli (Parfanovich and Sokolov, 1965). The perinuclear location of the first measles antigen to appear was confirmed in FL cells by Toyoshima et al. (1960b), the only difference from HEp_2 cells being that aggregates of antigen were irregularly shaped. In FL cells staining of measles virus antigen within nuclei was not seen so often, but in the few cells observed it did coincide with intranuclear inclusion bodies. The same sequence of events was again seen in FL cells by Karaki (1965a) who also found that specific fluorescence coincided with inclusion bodies in the paranuclear and perinuclear positions. Palacios (1965), examining measles virus-infected HeLa cells, not only saw specific fluorescent speckling of nuclei within 10 h of infection, but also states that 50 h later there was no specifically-stained antigen in nuclei. This is the only recorded instance with measles virus of the sort of transport of antigen from nucleus to cytoplasm which is characteristic of myxovirus infections. Unfortunately no details of the specificity of the antibody are available.

It remained for Norrby (1972) to re-examine the appearance of measles virus antigens in Vero cells and to identify them to some extent with antisera specific for the two main different components of the

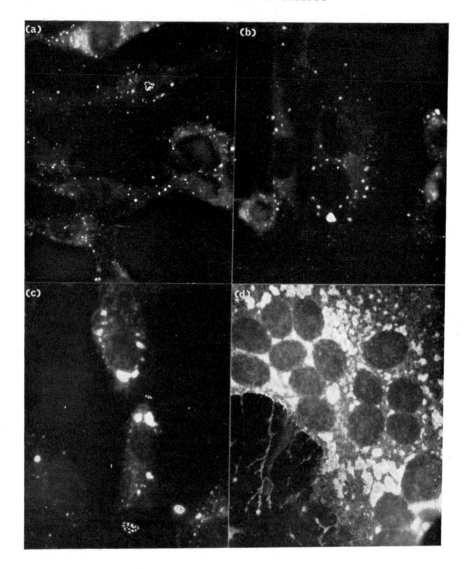

Fig. 6.2(A). HEp$_2$ cells infected with measles virus at a multiplicity $= 3$ and treated with guinea-pig anti-measles nucleocapsid, followed by fluorescein-conjugated anti-guinea-pig globulin. (a) 8 h after infection. Scattered cytoplasmic granules are specifically fluorescent. (b) 12 h. Some enlargement of granules. (c) 16 h. Fluorescent aggregates forming and faint, specific cytoplasmic fluorescence appearing. (d) 20 h. A syncytium showing typical fluorescent bodies and fluorescent nucleocapsid antigen in cytoplasmic processes.

virion the nucleocapsid and the haemagglutinin-haemolysin complex of the envelope (p. 54). He found that nucleocapsid antigen first appeared, from 8 h to 10 h after infection, as intracytoplasmic perinuclear dots, except in a few binucleate or multinucleate cells which had antigen within the nucleus. Nucleoli were surrounded by antigen which grew in bulk and tended to push nucleoli to one side as infection proceeded. A measles haemagglutinin-specific antibody stained cytoplasm also at 8–10 h after infection and the antigen appeared as a meshwork of fibres at one pole or other of the nucleus. The same serum began to stain cell membrane at 16 h to 18 h after infection, but membrane was not stained by treatment with anti-nucleocapsid serum. We have confirmed this work with guinea-pig monospecific antisera (Gharpure *et al.*, unpublished data; Fig. 6.2 A, B).

Thus appearance of antigen in the cytoplasm first and then in the nucleus seems to be the usual sequence of events in the replication of measles virus and it matches the order of appearance shown by eosinophilic inclusion bodies.

Protein synthesis. At present there is little information available concerning the synthesis of virus-specific protein in measles-infected cells. The time course of RNA and protein synthesis in measles-infected human amnion (AV3) cells has been reported by Portner and Bussell (1973). These investigators employed 6-azauridine and cycloheximide to inhibit the essential RNA and protein synthesis required for the production of infectious virus, salt-dependent agglutinin (SDA) and complement-fixing antigens. RNA and protein synthesis, essential for infectivity, was detected at 5 to 6 h and 6 to 8 h respectively and gradually increased until maximum levels were attained at 18 h. At this time, only 5% of progeny virus could be detected and in general, RNA and protein synthesis preceded the formation of infectious virus by at least 10 to 12 h. RNA synthesis, essential for the formation of SDA, was first detected 2 h after infection and reached a maximum level at 6 h, whereas protein synthesis, necessary for SDA formation, increased concurrently with the formation of antigens (both envelope and nucleocapsid), and was detected after 3 to 4 h. Protein synthesis commenced at 5 h and preceded that required for infectivity by 3 h.

Reversal of the inhibition of protein synthesis produced by cycloheximide treatment, by washing out the inhibitor from infected cells, revealed that the production of infectious virus was dependent on a period of "early" protein synthesis. This synthesis commences at 1 h after infection and was essentially completed by 3 h. The early protein(s) were synthesized prior to that essential for the production of CF antigen,

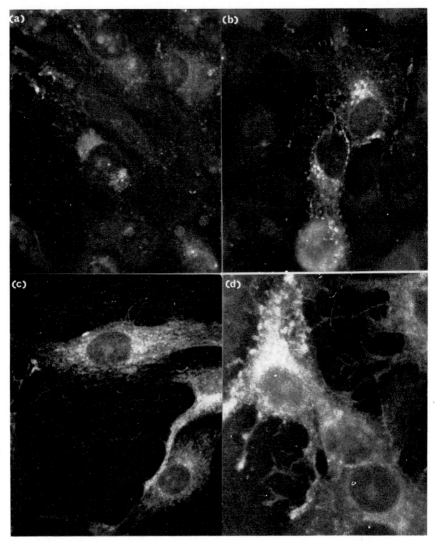

FIG. 6.2(B)   HEp$_2$ cells infected with measles virus at a multiplicity $= 3$ and treated with guinea-pig measles anti-haemagglutinin followed by fluorescein-conjugated anti-guinea-pig globulin. (a) 8 h after infection. Paranuclear granules and tufts of haemagglutinin revealed by fluorescence. (b) 12 h. Haemagglutinin still predominantly central but concentration is occurring at the cell membrane. (c) 16 h. Generalized specific cytoplasmic fluorescence and strong concentration at cell margins. No inclusion bodies or aggregates of antigen visible. (d) 20 h. Syncytium with much marginal and extruded antigen. No inclusion bodies visible.

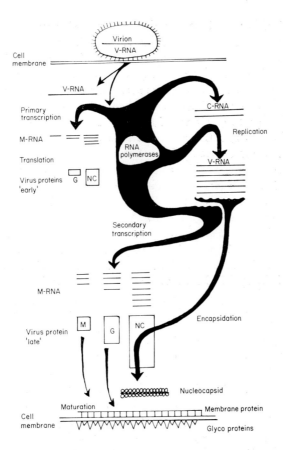

FIG. 6.3. Schematic representation of the probable events which occur during the replication of measles virus. V-RNA, = virion RNA; C-RNA = the strand complementary to V-RNA; mRNA = messenger RNA; NC = Nucleocapsid protein; G = surface, glycoproteins; M= membrane proteins. For sake of clarity, only the three main structural proteins are shown.

and SDA and infectious virus, and it was thought that they may be required for virus nucleic acid replication. It has been established that cycloheximide effectively inhibits the production of 50S progeny RNA in Sendai virus-infected cells, whereas production of messenger RNA is unaffected (Robinson, 1971). This result suggests that messenger RNA species were synthesized by the RNA polymerase carried into the cell by the virus, whereas the synthesis of 50S RNA is dependent on a different enzyme.

To sum up, it would appear that the proteins which are predominantly produced by messenger RNA, transcribed from the infecting virion RNA, are required for the transcription and replication of the 50S RNA genome and also the intact complementary strands. Messenger RNA molecules which are transcribed later are probably required for the production of structural components including the glycoproteins, nucleocapsid proteins and membrane proteins. There is also a considerable delay in the appearance of infectious virus after the synthesis of the RNA and antigens and it is likely that at least one of the virus-specified proteins is produced late in the growth cycle and may be involved in a rate-limiting step in the final maturation of the virions. There is some evidence from other enveloped viruses such as influenza virus (Lazarowitz et al., 1971; Mahy and Inglis, personal communication) and with Sendai virus (Lamb et al., 1976) that the membrane protein is produced later in the growth cycle and may play a regulatory role in determining the rate of virus maturation. A simplified scheme of the mechanism of replication is shown in Fig. 6.3.

MATURATION AND RELEASE OF MEASLES VIRUS

Data from various studies of the growth cycle of measles virus appear in Table 6.1 in which the appearance of infectivity marks the end of eclipse or the end of the latent period of infection, according to whether it is recorded as cell-associated or as free virus. Nothing is known about the detailed mechanisms of maturation and release of measles virus other than the morphogenetic studies which appear later in our text. The release process is continuous and probably slow (see Fig. 6.1 A, B) and very like that of paramyxoviruses (Duc Nguyen and Rosenblum, 1967; Howe et al., 1967). It should be remembered that the yield of infectious virus per cell is probably greater than it seems because of thermal inactivation of released virus. Successive 3-hourly or 8-hourly harvests have each been shown to yield as much virus as a single 24 h sample of supernatant culture fluid (Black, 1959b; Underwood, 1959).

Defective interfering particles produced during the replication of measles virus

Although the intact genome of measles virus is a single-stranded RNA molecule which sediments at 52S, and has a molecular weight of approximately 6.4×10^6, there have been a number of indications that populations of measles virus may contain RNA which sediments more slowly than the 52S component. Schleuderberg (1971) and Carter et al. (1973b) have shown that, like Sendai virus (Kingsbury et al., 1970), the types of RNA present in measles virus appear to be dependent on the multiplicity of infection of the input virus. Kiley *et al.* (1974) and Kiley and Payne (1974b) have extended these observations by demonstrating the presence of three distinct nucleocapsid species in measles-infected cells. Again, the types of nucleocapsids found are related to the passage history of the virus inoculum. Virus passed undiluted (UP) produces nucleocapsids which sediment at 110S, as well as 200S components, whereas only the intact genome is found when cells are infected with diluted passaged virus.

Hall *et al.* (1974) have prepared stocks of measles virus by passing it continuously at high dilution (10^{-3}) for a number of times and then passing in an undiluted manner for up to eight times. Virus was labelled with ^{3}H-uridine at passage numbers UP3 and UP8 as well as the DP virus.

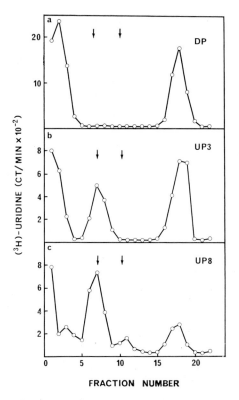

FIG. 6.4.   Sedimentation through 15–30% (w/v) sucrose gradients containing 0·5% (w/v) SDS of RNA isolated from virus released after (a) 10^{-3} diluted passage (DP), (b) three undiluted passages (UP3) and (c) eight undiluted passages (UP8). ○—○: ^{3}H-uridine, ct/min. (Hall *et al.*, 1974).

Sucrose gradient profiles of RNA isolated from virus particles prepared from UP3 and DP harvests showed that the predominant RNA present in DP virus was 52S, whereas the UP3 virus contained a species sedimenting at 18S. In UP8 virus there was a significant decrease in the amount of 52S RNA, and material sedimenting at 11S, 18S and 27S was present (Fig. 6.4). Virus harvests obtained after infection with UP3 virus were concentrated by forced dialysis and sedimented on discontinuous gradients. The virus bands were pooled and sedimented on linear 15–30% (w/w) sucrose gradients on a cushion of 60% sucrose. Most of the virus sedimented heterogeneously across the gradient, but a significant proportion of, presumably, aggregated material landed at the 60% sucrose cushion. Repeated sedimentation on sucrose gradients of regions A and B of the first gradient led to the separation of two homogeneous virus populations. The rapidly sedimenting species contained encapsidated 52S RNA and was infectious virus. In contrast, the slowly sedimenting species contained encapsidated 18S RNA and was non-infectious (Fig. 6.4). These sub-genomic particles were able to interfere with the replication of infectious virus. The interfering activity was destroyed by UV irradiation suggesting that the 18S RNA was the biologically active component (Table 6.2).

TABLE 6.2

Interference by measles virus containing "18S" RNA with plaque formation by measles virus containing "52S" RNA genomes

HAU of "18S" RNA Virus Added	Average Number of Plaques
0	269
256	87
25·6	152
2·56	228
256 (UV irradiated)	272

Virus containing 52S RNA inoculated at a 10^{-5} dilution. UV irradiation was carried out using a BTL Ultra-Violet lamp employing two strip lamps at 25 37Å and 35 60Å respectively. Irradiation was continued for 20 min at room temperature. (Data from Hall *et al.*, 1974).

This type of interference would account for the effect of continued passage of undiluted virus which produced a non-syncytial type of infection and lower yields of infectious virus. The findings would also account for the differences in the type of virus products released and the types of RNA produced in infected cells by UP and DP variants (Norrby *et al.*, 1970; Parfanovich *et al.*, 1971).

It would appear, therefore, that measles virus does not differ from VSV, Sendai virus or NDV in its ability to produce defective interfering particles. However, at present, we do not know the exact relationship of the sub-genomic RNA in the interfering component to intact genomic RNA. It has still to be established whether this is of the same sense or complementary to the virion RNA. It is very probable that in any preparation both types of RNA strands are present. Also, since it is probable that some of the large particles contain multiple copies of nucleocapsids, it is very likely that at least a few particles contain a single intact genome plus a number of sub-genomic interfering units. At present, the results of infection of a cell with such a particle are not known, but it would appear to offer a situation within a single cell similar to that resulting from infecting monolayer cultures with UP virus. It would seem to be important to study the role of these complex virus particles in viral pathogenesis and to establish what relationship, if any, they may have to the genesis of persistent infections.

MORPHOGENESIS OF MEASLES VIRUS

No very exact picture has been drawn of the sequence of intra-cellular changes which might be interpreted as the steps by which measles virus is built. One cause of this is the difficulty in obtaining synchronous growth in a sufficient number of cells to give a good average representation of what is going on at any one time in the growth cycle. Another may be the fairly diffuse scattering of maturation sites on the infected cell surface. In spite of these drawbacks the growth process can be classed along with that of paramyxoviruses with which the measles virus has in common a tubular nucleocapsid, an envelope with inserted spikes of one or more kinds and maturation at the cell surface (Waterson et al., 1961; Waterson, 1965). The types of inclusion found and the stages of morphogenesis in tissue culture are illustrated in Figs. 6.5, 6.6, 6.7, and 6.8.

The first components of measles virus to be recognized for certain in the cell were strand-like structures which were most likely to have been nucleocapsid. This was achieved by electron microscopy of inclusion bodies which had first been precisely identified in infected HeLa cells by conventional microscopy and photography (Kallman et al., 1959). The strands, which had a diameter of about 20 nm, were loosely arranged with occasional groups in line, and continuity could be seen in places between the random and the parallel arrangements. They were also described, again with Edmonston virus in HeLa cells, by Tawara et al. (1961).

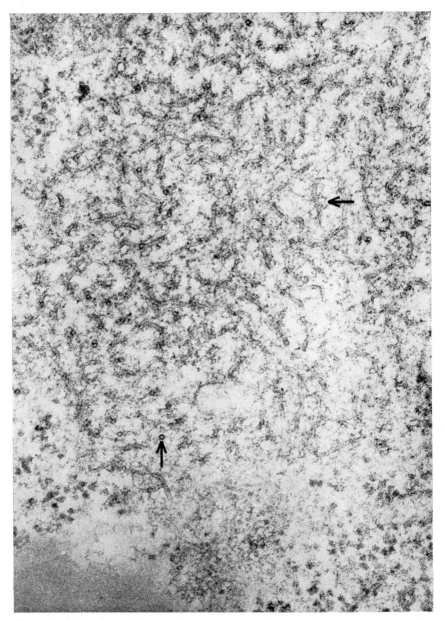

Fig. 6.5. Measles nucleocapsid aggregate in the cytoplasm of a persistently infected HEp_2 cell. Individual capsids are hollow in cross section and in longitudinal section show a striated pattern. This type of nucleocapsid has been described as "fuzzy" (see text) because of the dense strands which radiate from it (arrows). Magnification $\times 90,000$.

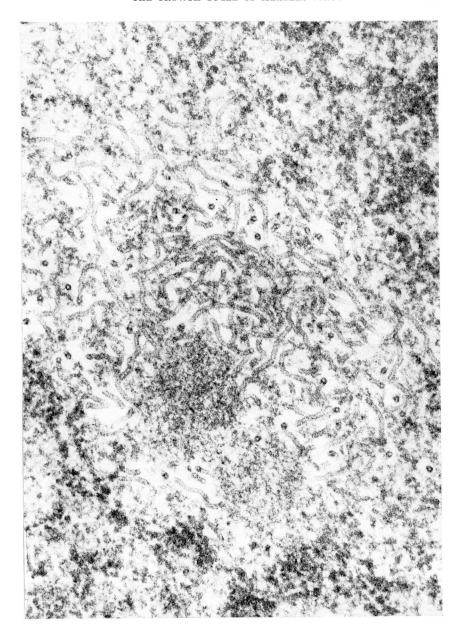

FIG. 6.6.   Smooth nucleocapsids (see text) within the nucleus of an infected
Vero cell. Magnification ×90,000.

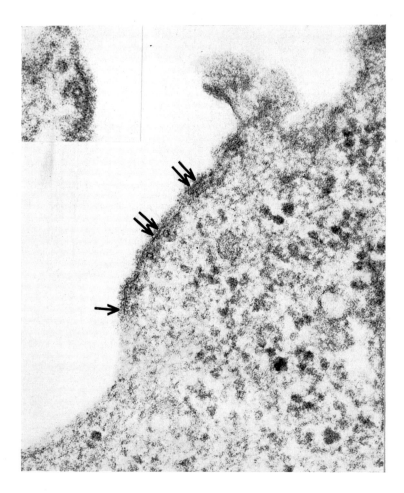

Fig. 6.7.   Nucleocapsid, surrounded by an electron-dense material, is aligned
immediately beneath the cell plasma membrane which has itself been
altered. This micrograph illustrates the deposition of a dense material
underneath a membrane which retains its tri-partite appearance (single
arrow); then, as spikes are seen on the outer surface, the tri-partite
structure is altered to become a single dense line (double arrows). Magnification ×120,000. *Inset:* Detail of the altered membrane structure,
spikes and underlying nucleocapsid surrounded by dense material. Magnification ×160,000.

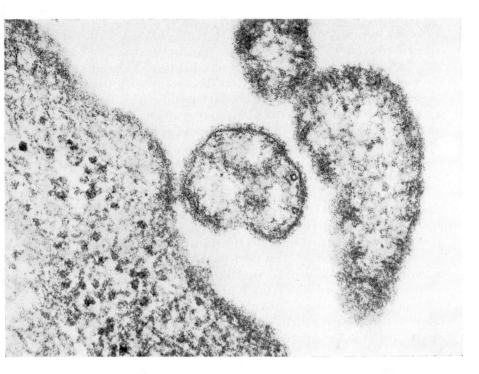

Fig. 6.8.   A complete measles virion budded from the host cell. The envelope is seen as a single dense line bearing a fringe of spikes on the outer surface. The nucleocapsid is seen at the periphery and traversing the central region of the particle. Magnification ×120,000.

These electron-microscopic changes were identifiable in dog-kidney cells infected with Edmonston virus two days before conventional inclusion bodies became visible in the cytoplasm at the 3rd or 4th day, or 2 to 4 days before intranuclear inclusion bodies became visible, from the 7th to 10th day after infection. Strand-like formation was apt to mark the centre of the inclusion material and the strands were recognized as tubular, the external diameter being 15 to 20 nm, as compared with 5 to 10 nm for the internal diameter (Tawara, 1964, 1965).

Mannweiler (1965) examined the Edmonston and Pittsburgh strains of measles virus, infecting HeLa cells, and described the formation of virions by the protrusion from the cell-membrane of buds, into which tubules from the cytoplasm were enfolded. He reported the occurrence of two kinds of particle, one kind electron-dense and about 150 to 220 nm in diameter the other more electron-translucent and of much larger size, about 400 nm in diameter. This led him to speculate whether or not the larger kind of particle might incorporate so much host-cell membranes as to cause autoimmune responses.

Even greater differences in particle size were given by Matsumoto (1966) who added the opinion that the membranes of the smaller particles, 200 nm in diameter, were denser than the membrane of larger particles which had a diameter up to 700 nm. Labelling with ferritin-labelled measles-specific antibody resulted in heavy coating of the small spherical particles and also to coating of the envelope of the infected cell. Matsumoto, who was examining measles virus-infected Vero cells at the 5th and 7th days of incubation, when giant cells and inclusion bodies had been formed, noted that parallel arrays of tubules were characteristic of nuclear inclusion bodies, but not of cytoplasmic inclusions and he also observed that there seemed to be a decided difference between the two sets of tubules, those in the cytoplasm having many up to 30 nm across as compared to 15 nm for the diameter of intranuclear tubules. This dimension has since been used to "identify" intranuclear, tubular material in brain cells derived from the pathological material of subacute sclerosing panencephalitis (Tellez-Nagel and Harter, 1966).

Some sort of time sequence was obtained by studying HeLa cells, infected with Edmonston measles, at intervals, 18 to 20 h, 30 to 42 h and 96 to 120 h after infection (Nakai and Imagawa, 1969), and in monkey kidney cells from two species of monkey at 15, 20, 30, 40 and 50 h after infection (Nakai et al., 1969). In the HeLa cell experiments the dimensions of tubules are given as less than previous measurements, being 15 to 17 nm within virus particles and in intranuclear inclusions. At high multiplicity of infection they were seen in HeLa cells at 18 to 20 h after inoculation. Thickening of the cell-membrane followed at 30 to 42 h after infection and strands collected internally contiguous to the thickened membrane which differentiated into fringe-like structures. Budding which involved the fringed membrane and incorporated tubular strands from the cytoplasm was also seen then. Intranuclear tubules were not seen until 96 to 120 h after infection and appeared within the bounds of intact nuclear membranes.

In BS-C-1 monkey cells, bundles of fibrils within which were scattered tubules were seen at 30 h after infection. The cytoplasmic material increased, and tubules, but no fibrils, next appeared in the nucleus. Budding virus particles, only some of which contained tubules, were not seen until 65 h after infection. In BS-C-1 cells excessive multiplicity of infection (m = 10) caused tubules to become visible first in the nucleus at 22 h after infection then in the cytoplasm at 36 h, a sequence which also characterized the growth of virus in Rhesus monkey kidney cells at a lesser multiplicity of infection. When a dilute passaged virus was used, which caused a rapid cytopathic effect and resulted in high yields of infectious virus ($10^{7 \cdot 0}$ p.f.u./ml) tubules were seen in the cytoplasm at 18 h and budding took place after 30 h of incubation.

These observations probably give the correct order of events but the delay in finding characteristic changes as compared with the shorter durations of the measles eclipse phase at similar multiplicities (Table 6.1) probably indicates mainly the time required for components of the virus to grow to a concentration sufficient to provide representative appearances.

Immunological identification of virus antigen in the inclusion bodies of measles virus-infected cells, which has been noted in studies of virus synthesis by the fluorescent antibody method (p. 81), was obtained by using peroxidase labelled antibody, a method which can be applied also to electron microscopy. By light microscopy both intranuclear and intra-cytoplasmic inclusion bodies became labelled, but nothing was traced at the cell surface (Boyle and Atherton, 1971). A more detailed account of a subacute sclerosing panencephalitis strain of measles virus, Mc-Lennan, was given by Dubois-Dalcq and Barbosa (1973) again by electron microscopy. Patches of virus antigen were identified at the surface of infected Vero cells and these patches were usually related to under-lying tubular or capsid material. Of two types of tubule seen, fuzzy or smooth, only the fuzzy tubule became labelled and inclusions stained variably, that is they were either labelled or unlabelled, but not patchy, like staining of the cell-membrane.

Maturation of measles virus in nervous tissue seems not to be different from that in cultured renal cells. Hamster dorsal root ganglion, in culture, is susceptible to long-term infection by measles virus and shows by electron microscopy that fuzzy nucleocapsids form a dense inner lining of parts of the cell membrane at which budding takes place (Raine *et al.*, 1971). In this article the dimensions of the nucleocapsid correspond more closely to the large diameter given by Matsumoto (p. 94), being 20 nm in diameter without its coating material and 40 to 50 nm with it. When, at 18 days post-inoculation, nuclear changes were eventually seen, it was noted that the intranuclear capsid material was uncoated and had a diameter of 20 nm. An earlier report indicated that in long-term infection and late in short-term cytocidal infection there was a relative lack of capsid material in budding virions (Raine *et al.*, 1969). This was not the case in the second study.

The morphogenesis of strains of measles virus derived from patients with subacute sclerosing panencephalitis have also been examined in or-gan culture (Raine *et al.*, 1973) and in hamster brain *in vivo* (Raine *et al.*, 1974b; Dubois-Dalcq *et al.*, 1975) as well as in tissue cultures of kidney cells from *Cercopitheus aethiops* (Draganescu *et al.*, 1973). Narrow smooth and wide fuzzy tubules were seen by all observers, but one strain revealed a difference from wild-type virus (Raine *et al.*, 1973). Whereas the nucleocapsids of wild type virus are aligned under the cell membrane in association with fringed material and with budding, the subacute sclerosing panencephalitis strain, Mantooth, did not show nucleocapsids aligned at cell membrane and the buds contained no tubules. Comparisons need to be made between Mantooth and other strains of subacute sclerosing panencephalitis virus and also with non-yielding carrier cultures of measles virus (p. 104).

Note added in proof. Recent work with fluorescent antibody prepared against the 36,000 MW virus membrane protein points to a special association with nucleocapsid antigen and also with morphogenesis of the virion. The nucleocapsid and membrane protein antigens are always congruent in the cytoplasmic inclusion bodies and are the first to be detected in the cell at 2–4 h after infection and about 4 to 6 h before haemagglutinin and haemolysin are seen. Thereafter both are transported to the membrane of the infected cell where the nucleocapsid remains internal but the antigen of membrane protein is easily traceable on to the external surface of the cell membrane along with haemagglutinin and haemolysin.

Gel-diffusion studies with mono-specific immune sera and immuno-electron-microscopy have confirmed these relationships. (Fraser *et al.*, 1977).

Persistent infection by measles virus

INTERPRETATION: (1) *The ability to initiate persistent infection* in vivo *and* in vitro *is an important characteristic of measles virus.* (2) *The operative mechanisms are not understood but inhibition of assembly or incomplete maturation are striking features.* (3) *The genesis of persistent infection lies most likely in the production of defective interfering particles.* (4) *Abortive infection in the central nervous system has also been noted.*

BIOLOGICAL ASPECTS OF PERSISTENT INFECTION

In vivo *persistence*

Carriers of latent measles virus were first detected by the isolation of virus from cultures of apparently normal monkey kidney in which spontaneous vacuolation frequently appeared. About 57% of 99 specimens yielded a foamy agent, but measles virus was recognized separate from or along with it by the existence of intranuclear inclusions. All the intranuclear inclusion agents from 17% of the same specimens were serologically identical with measles (Ruckle, 1958). Such findings did not determine the duration of persistent infection *in vivo*, but two later studies showed measles virus in monkey spleen $2\frac{1}{2}$ months after infection (Ruckle-Enders, 1962) and in the brain of monkeys that had been inoculated intrathalamically with Edmonston virus 5 months before and had remained symptom free until killed (Albrecht *et al.*, 1972).

One claim to have isolated measles virus from healthy human adults is credible. Enders-Ruckle, in discussing the form which immunity to measles in human beings may take, mentioned recovering two strains of virus from lymph node cultures (Enders-Ruckle, 1965). The fact that haemadsorption of infected cells was not present at the first few passes and haemagglutinin production tended to remain low, even in later passes, suggests that these were fresh isolates and not contaminating laboratory organisms.

The chronic infection of subacute sclerosing panencephalitis will be discussed in Chapter 9 but prolonged measles infection has been found in other organs than the brain. Deep ulcers in the ileum, granulomata in lung and liver, and giant cells in bowel-wall, lung and all gut-dependent areas, were found in a child who contracted measles at the age of thirteen months and who died nine months later. There was evidence of lymphocytic depletion. Thymic dysplasia was present with macrophages in the thymus, but no characteristic measles lesions were seen there. There was also an excessive amount of circulating immunoglobulin M but it did not represent measles-specific antibody (Marsden, 1973).

In vitro *persistence*

Rustigian (1962) first demonstrated persistent infection by measles virus *in vitro*, using HeLa cells, at a time when his experiments seemed to have no clinical significance. Its importance has been realized since subacute sclerosing panencephalitis came to be considered as a measles virus infection of long duration, perhaps dating from the primary attack of measles.

About the year 1962 also it came to be realized that certain enveloped RNA viruses can establish a type of transmissible balanced infection, more or less free of cellular degeneration, in which most cells produce virus antigen. Little or no infectious virus may be formed and the morphology, plating efficiency and growth rates of infected cells are indistinguishable from those of uninfected cells. Measles virus and parainfluenza viruses easily form this sort of relationship so that infected cells are usually detected by haemadsorption or by fluorescent antibody tracing of virus antigen. High multiplicity of infection plays an important part in establishing and perhaps in maintaining persistent intracellular infection, but interferon does not, and it is now accepted, especially with rhabdo viruses, that incomplete virus particles form the principal auto-interfering agents. Superinfection by the same virus is not usually possible, but superinfection by unrelated viruses is interfered with only very slightly (Walker, 1964).

Virus-yielding carrier lines. One successful method of initiating such cultures is to cultivate cells surviving a heavy infection. The survivors may have been protected from full cytocidal activity of the virus by interferon or by interfering particles (DI particles, p. 86). When the balance between replication and destruction is difficult to stabilize, virus antibody may be applied to the culture during successive passes. Rustigian (1962) made use of human immunoglobulin in the growth medium during the genesis of persistently infected HeLa cells, the antibody having, of course, no action against virus passed from parent to daughter cells during mitosis. Further studies by Rustigian showed that clones of cells from the infected culture could be virus free, but most were infected and only ceased to show some cytopathic effect very gradually as passage continued. Cells bearing typical intranuclear inclusion bodies declined in number to not less than 0.2% per culture but virus antigen was present in nearly all cells. An inoculum of $10^{5.0}$ $TCID_{50}$ of measles virus had no effect in such cultures, but vaccinia virus, herpes simplex virus, poliovirus and syncytial foamy agent caused the same cytopathic changes as they did in normal HeLa cell layers. Application of actinomycin D did not affect the growth of the culture nor alter the variable pattern and amount of fluorescent antibody staining seen in different cells, which Rustigian supposed to be due to individual host-cell factors (Rustigian, 1966a). By selecting clones and growing the cells in the presence of measles antibody a persistently infected sub-clone was obtained which produced practically no infectious virus (Rustigian, 1966b).

A greatly reduced incidence and intensity of haemadsorption was noted, accompanied by a very much reduced amount of virus antigen at the cell surface, as revealed by fluorescent antibody. No free haem-agglutinin was detected and the yield of infectious virus was only $0.8–0.9$ TCD_{50} per ml of culture. The cell cytoplasm also seemed to be unstained by fluorescent antibody, antigen being confined to cytoplasmic aggregates of virus and in many cells intranuclear aggregates too were only faintly stained.

Since then, similar cultures have been established in HeLa cells, KB cells, lung cells, hamster kidney cells, African green monkey cells, BS-C-1 cells, mouse brain cells and hamster brain cells, some of these of

TABLE 7.1
Persistent infection by measles virus

Reference	Cell system	Virus strain	Method of establishing	Presence of CPE
Rustigian 1962–66	Human HeLa	Edmonston	Survivors of marked CPE	None
Norrby 1967	Lu 106	Edmonston	Survivors of low multiplicity infection.	Present
Minagawa 1971a,b.	Human HeLa	Toyoshima	Survivors of infection Unstable for several months.	?
Knight et al., 1972–74	Hamster embryo fibroblast	Edmonston-Schwarz (attenuated)	Survivors of low multiplicity infection.	?
Gould 1974–75.	Human HEp$_2$	Edmonston	Survivors of marked CPE following low multiplicity	None
Burnstein et al., 1974.	Monkey BS-C-1	SSPE strain	Cell fusion and fluctuating CPE	1 line stable 1 line cyclical syncytia
Menna et al., 1975a,b.	Monkey BGM	Edmonston mouse-adapted	Co-cultivation	?
Zhdanov and Parfanovich 1974.	Chick embryo fibroblast	Moscow M6	m = 0.01 TCD$_{50}$	None

Reference	Yield of virus	Percentage of antigen bearing cells	Haemadsorption	Reactivation	Interference with other viruses
Rustigian 1962–66	<1 TCD$_{50}$ per 10^4 cells	85% to 95% surface antigen 0·02%–0·2%	Reduced in non-yielders	N.D.	None, but refractory to measles virus
Norrby 1967	Low $10^{1.5}$–$10^{3.0}$ TCD$_{50}$ Rises at 33°C	?	?	No effect Actinomycin D. Increased yield at low temp	?
Minagawa 1971a,b	9% yielder cells	?	N.D.	Carrier type reverts after pass in Vero.	None to mixed inoculum
Knight et al., 1972–74	1 pfu per 20,000 cells at pass 20 × Rises at 33°C	? All	N.D.	Yes by co-cultivation and plating cells	N.D.
Gould 1974–75	10^2–10^3 pfu/ml (decreasing with passage)	All Surface stain on > 90% cells	50%	Actinomycin D increases yield 1 log	None Refractory to measles
Burnstein et al., 1974	None None	All cells Syncytia only	None Slight	None None except in brain	N.D.
Menna et al., 1975a,b	None	80%–90%	< 1%	None	None, but partially refractory to measles
Zhdanov and Parfanovich 1974	None CAV 10^2 TCD$_{50}$	10–15%	N.D.	N.D.	N.D.

relatively short duration, but all low-yielders or non-yielders of infectious virus, all showing minimal cytopathic changes and most of them
displaying some or all of the observations first recorded so thoroughly by
Rustigian (see Table 7.1).

Minagawa (1971a) set up and studied a more active persistent infection
in HeLa cells, using the "cell-survivor" method to establish the culture.
He measured virus-releasing cells and found that the number fluctuated
from 11% to 52% of the cell number in a cloned carrier culture
which was derived from a line that multiplied at the same rate as uninfected cells up to six days after plating, but which had itself a superior
plating efficiency. The average yield of virus per cell was 8% to 9% in
$TCID_{50}$. Non-releasing cells, recovered from the carrier culture, had
the same susceptibility to superinfection by measles virus as uninfected
HeLa cells.

Minagawa (1971b) re-isolated and examined the carrier virus and noted
that in HeLa cells it had a slower growth rate, lesser yield and lesser
cytopathic effect than had the starter virus. Inoculated into Vero cells,
it grew only slowly at first, but in four days reached the same final titre
as the original stock measles virus could produce. Plaques of original
and carried virus were alike in size and shape, but those of carried virus
grew more slowly and were ten to a hundred times fewer.

The ability of carrier virus to set up a carrier state when inoculated
into fresh HeLa cells was lost by one passage in Vero cells and normal
plaque forming ability was restored. Unfortunately, the effect of a single
cycle of growth in Vero cells does not seem to have been tested.

Norrby's carrier line in Lu 106 cells was set up at a multiplicity of 1
$TCID_{50}/10$ cells and again survivors began to repopulate the culture flask.
The carrier line did not grow so well as the original cells, carried 95%
of cells showing specific haemadsorption and yielded haemagglutinin
only two to four-fold less in amount than did stock measles in lytic
infection. Nevertheless, yields of infectious virus were 10^4 to 10^5 times
less in the carrier culture. Vaccinia virus, Sendai virus and adenovirus
were not interfered with in the carrier line, but polio virus plaques were
somewhat reduced. Neither the cytopathology nor the virus-yield of the
carrier state was affected by actinomycin D at a concentration of
0.05 μg/ml. However, the balance was upset at lower temperatures of
incubation, syncytial formation being 10 to 60 times more evident at
$33°C$ than at $37°C$. The ratio of infectious virus to HA also rose eight-fold
(Norrby, 1967).

Gould established a chronic carrier state in HEp_2 cells which has not
been fully characterized, but low yields of virus were characteristic.
There was a steadily decreasing amount of membrane-bound antigen on

the exterior of the cell, cytoplasmic inclusions were abundant and, unlike Rustigian's HeLa cell cultures (Rustigian, 1966b), the amount of measles antigen to be found in nuclei increased greatly after two years of continuous culture. The pattern of these nuclear inclusions is now extremely regular and characteristic (Fig. 9.9), and in size some approach the large intranuclear central masses seen in neurones of subacute sclerosing panencephalitis. Gould also found that his carrier virus had altered, tending to set up the carrier state rather than a lytic cycle, in HEp_2 cells; the virus growth was temperature-sensitive and the proportion of miniature plaques found in infected Vero cells as compared with the starter strain of virus was greatly increased. The carried virus could be bred true at limit dilution in the sense that when plated out the virus displayed a ratio in the count of miniature to large plaques that stayed constant at not less than 100 to 1 (Gould, 1974).

The Schwarz strain of measles, inoculated into hamster embryo fibroblasts (HEF) at a multiplicity of 0·01 and, after 21 days incubation, passed five times more, set up, from surviving cells, yet another carrier state (Knight et al., 1972). The yield of infectious virus was steady at 1 p.f.u./20,000 cells after 20 passes, but incubation at 33°C raised this to 1 p.f.u. per 200 cells. On incubation at 39°C, virus synthesis was reduced to the appearance of a little perinuclear antigen only. However, one in every ten hamster embryo fibroblasts could act as an infectious centre and co-cultivation with BS-C-1 cells released virus to a titre of $10^{6·0}$ p.f.u./ml and was accompanied by cell destruction in six hours. Several attributes of the system have been examined at different times. One of particular interest was the effect of cell fusion on the intracellular location of measles antigen. From being dispersed in the cytoplasm, it became compactly perinuclear in position after cell fusion and 12 or 16 h later migrated centrifugally from the nucleus. Degeneration of cells followed in 4 h. This antigen was identified in the electron microscope as, or being congruent with, nucleocapsid. Whole BS-C-1 cells were required for this sequence of events to take place. It was stopped by carbenzoxytripeptide which also stopped the growth of the latent virus (Knight et al., 1973).

Exposure of these mixed cultures to RNA inhibitors (azacytidine) or protein inhibitors (cycloheximide) did not prevent or delay the onset of CPE or release of infectious virus. It would appear that progeny RNA and protein were present in the latently infected HEF cells and that no new synthesis of protein or RNA was required prior to maturation of infectious virus. This implied that a "factor" was added by addition of another susceptible cell. The quick release of virus upon contact between the two cell types, the presence of large amounts of measles virus-specific antigens in the cells, the existence of preformed progeny RNA and the need for little or no new protein synthesis for early release after co-cultivation suggests in addition that a late maturation step (perhaps involving a budding process) is the site of the block in the synthesis of infectious virus.

The nature of the virus in this carrier culture was investigated also (Haspel et al., 1973), and it proved to be temperature-sensitive. Interesting differences were noted between virus released by incubation at low temperature (TR) and virus released by co-cultivation (CC). The co-cultivated virus, like the Schwarz strain of starter virus, grew better at 33·5°C than 39°C. One other property of the parental virus was shared by released (CC) virus, namely the ability to grow better in multiplying hamster embryo fibroblast (HEF) cells than in stationary cells. The heat-resistance of the released viruses was alike and TR virus produced small plaques, CC virus large plaques. However, plaque size varied independently of temperature sensitivity for CC virus also bred small plaque-producing viruses that were not temperature-sensitive. After many subcultures, stage 3 of the author's study, changes in that property of released virus took place. TR virus was released in greatest amounts at lower temperatures of incubation but was no longer temperature-sensitive. Titres were higher than in stage 2 and released virus produced larger plaques. This is a dangerously condensed résumé of an important paper, but it is clear that the measles virus in artificially produced persistent infection is no more static in its properties than strains grown from subacute sclerosing panencephalitis (p. 146).

Non-yielding carrier lines. Experiments on neurotropism of measles virus (Chapter 8) in which protracted infection is a relatively common event, especially in partially immune or older animals, has led to the establishment in tissue culture of viruses derived from such sources. A system devised by Burnstein employed BS-C-1 cells and brain cells cultivated from a biopsy specimen; the patient was 14 years old and had subacute sclerosing panencephalitis. The system showed particularly interesting cycles of syncytial formation, followed by tissue break-down and healing and eventually two non-yielding lines were set up, one (IP-3) forming syncytia and one (IP-3-Ca) of healthy appearance. Culture IP-3-Ca showed haemadsorption of slight intensity; both cultures synthesized measles antigen, IP-3 in syncytia only and IP-3-Ca in all cells; both caused encephalitis on intra-cerebral inoculation into young (weanling) hamsters; and both gave CPE on co-cultivation with BS-C-1 cells. It is most interesting that although one culture caused syncytial formation and massive cell breakdown on co-cultivation, no infectious virus was released. In this last respect it differed from all other measles carrier systems yet described,* but the fact that occasional recovery of virus followed intra-cerebral inoculation indicates that the complete genome had not altogether vanished from the persistently infected culture (Burnstein et al., 1974).

The latest carrier line to be described also used a virus of neurotropic origin, Adams' and Imagawa's mouse-adapted strain of Edmonston virus that had been carried in C3H mice before being inoculated into cultured mouse-brain cells. Co-cultivation was carried out primarily with a stable line, BGM, of African green monkey kidney cells and, without further addition of measles-infected brain cells, the culture was maintained for 260 days. Weekly passage was then begun. For 20 days there was a yield of 10–200 p.f.u./ml. Thereafter none was found when tested at intervals up to 750 days. There was no production of haemagglutinin—less than 1% of cells showed measles-specific haemadsorption—and at the 15th pass 80–90% of the cells showed intracellular fluorescent antibody staining which was confined to the cytoplasm; all cells had measles antigen at the membrane. At passage 26, that is the 440th day of culture, less than one cell in a hundred had antigen at the membrane. Although activation of virus by drugs and by co-cultivation was not possible, actinomycin D, colchicine, bromodeoxyuridine and a few other drugs increased the number of cells giving measles-specific haemadsorption: enucleation of cells by cytochalasin had the same result, 30 to 40% of cells becoming positive to haemadsorption as compared with the 70% produced by actinomycin D (Menna et al., 1975a).

This system, like the previous one, did not yield virus on co-cultivation with monkey kidney cells and it is the second system known to have behaved in this way. The culture was only partially refractory to superinfection with measles virus and fully susceptible to mumps virus and vesicular stomatitis virus. Interferon failed to convert the culture to an antiviral condition (Menna et al., 1975b).

Long-term infection in "brain" cells

Brain cells in culture are susceptible to infection (Table 7.2). Gibson and Bell (1972) noted that infected explants of one to three-day-old mouse brain grow better than non-infected ones. Virus antigen, traced by fluorescent antibody, was cytoplasmic and perinuclear in situation, not intranuclear nor situated at the cell membrane, and was present in only a few cells. The yield of virus was very low 0·5 to 2·0 $TCID_{50}$ per 0·1/ml. The absence of specific cytotoxicity, using guinea-pig complement and rabbit anti-measles serum, was consistent with the absence of measles virus antigen at the surface of infected cells.

* Rustigian's non-yielding clones (1966b) preceded the use of co-cultivation and cell fusion.

TABLE 7.2.

Prolonged infection in brain cells

Reference	Cell system or host	Virus strain	Method of establishing	Presence of CPE
Gibson and Bell, 1972	Mouse brain	Mouse adapted (Rapp)	Added free virus	?
Bather *et al.*, 1973	Monkey brain	Edmonston Schwarz	Free virus Persisting	Present
Wear and Rapp, 1971	Hamster brain *in vivo*	Edmonston Schwarz	Intra-cerebral	NA

Reference	Yield of virus	Percentage of antigen-bearing cells	Haemadsorption	Reactivation	Interference
Gibson and Bell, 1972	0·5–2·0 TCD_{50} per 0·1 ml.	Very small	?	ND	ND
Bather *et al.*, 1973	Low	90%	Decreases with each pass	?	?
Wear and Rapp, 1971	By co-culture only Very low	NA	NA	NA	NA

NA = Not available; ND = Not done.

A rather longer persistence in brain cells, virtually a carrier state, has been described by Bather et al. (1973). The substrate was a diploid cell line derived from the brain of an African green monkey, infected at the 10th pass, and still 95% euploid at the 35th pass. The inoculum was the Schwarz derivative of Edmonston vaccine. Some initial cytopathic effect was followed by fluctuating CPE which affected less than 25% of the cell number, and was accompanied by only a low yield of virus at each pass. Nevertheless 85% to 95% of cells contained measles antigen, as seen by immunofluorescence, and the number of cells scoring as infectious centres decreased gradually with continued passage. Unfortunately, figures for only three passes are shown.

The Schwarz strain of measles virus can be quite lethal to newborn hamsters when inoculated into the brain, but when the hamsters are born to mothers which have been hyper-immunized against measles, a prolonged intracerebral infection ensues in many of the survivors of infection. Symptoms of encephalitis have been recorded as appearing at 70 days after the initial infection. Induction of symptoms by cyclophosphamide immunosuppression was accompanied by growth of the virus in three out of three cases tested (Wear and Rapp, 1971).

MOLECULAR ASPECTS OF PERSISTENT INFECTION

RNA synthesis in persistently infected cells

Very little information is available on the biochemistry of cells persistently infected with measles virus and although both electron microscopic and immunofluorescent evidence suggests that the cells contain considerable amounts of virus antigen there are no hard facts about the amounts or types of virus specific RNAs or proteins that are present in persistently infected cells.

Winston et al. (1973) have investigated the RNA synthesized in HeLa cells persistently infected with measles virus and found that the virus-specific RNA was qualitatively similar to the RNA in lytically infected cultures when cells were labelled with ^{3}H-uridine in the presence of actinomycin D. However, comparison with the virus-specific RNA synthesis in primary and persistently infected HeLa cells showed that the primary infection yields three times more of the low-molecular weight RNA species, mostly 18S RNA. This observation is similar to those of Blair and Robinson (1968) for other paramyxovirus infections where the ratio of low to high molecular weight species is higher in avirulent or mildly virulent infections than in virulent infections. At present, no information is available on the details of the types of RNA present in the 18S peak isolated from persistently infected cells. It is possible that some of the mRNA components are absent or greatly reduced. However, as labelling patterns of RNA result from both the rates of synthesis and degradation of the species formed, it is unlikely that such a comparison can give a true insight into the situation in persistently infected cells. Furthermore, as mentioned on p. 80, it is becoming more and more evident that the use of actinomycin may not be without effect on either the synthesis or stability of measles 50S RNA and perhaps also on the half-life and size of mRNA. Schluederberg *et al.* (1972) have reported the apparent inhibition of 50S RNA in cells lytically infected with measles virus and we have been consistently unsuccessful in numerous attempts to label the RNA isotopically in a number of persistently infected cultures in the presence of actinomycin D. Hence, it is important to interpret with caution the results of actinomycin experiments since the profiles obtained may not be a true reflection of what is occurring in the non-treated cells. Further, it is essential to know the precise morphological state of cultures during such investigations, to ensure that cells are not passing through a period of crisis, or that a transient lysis is not being induced by the manipulations involved in the labelling experiments. For example, a change in medium may result in the transient onset of a lytic crisis (ter Meulen and Martin, 1976) and it is well known that actinomycin D can effect an increase in the yield of infectious virus released from some persistent cultures.

Evidence for the integration of measles genome

Recently, a number of reports have suggested that reverse transcription and integration of RNA virus genomes into the genome of the host cell may be mediated by enzymes from oncornaviruses and that this may play a role in the onset and maintenance of persistent infections. Zhdanov (1975) has shown by RNA–DNA hydridization and transfection studies that cells persistently infected with measles virus contain in their genomes DNA sequences complementary to the RNA sequences of the virus. Furthermore, Zhdanov showed that the DNA of leukocytes and tissues from patients suffering from systemic lupus erythematosis contained homologous sequences to measles RNA while no such sequences were detected in DNA from normal human tissues, nor in DNA from the leukocytes of measles patients taken at an early stage in the disease and in convalescence. Simpson and Iinuma (1975) have also found evidence for the integration of RNA from RS virus and successful transfection experiments showed that complete genomes were integrated. Zhdanov proposed that during the genesis of persistent infection the measles or measles-like RNA integrates after interaction with a latent oncornavirus, resulting in the reverse transcription of the RNA into the DNA of the host cell. The integrated viral RNA may be translated and this alters the antigenic properties of the cell. The possibility that such virus–virus–cell interaction exists is of much current interest, especially as it would explain many of the features of long-term persistence. However, it remains to be seen whether such interactions are essential for the genesis of persistent infection or whether successful integration is an incidental occurrence, usually accompanying a protracted infection and therefore unrelated to either maintenance of a persistent state or to disease processes. Nonetheless, many laboratories that have access to neurological and other tissue containing measles antigens will no doubt be actively searching for integrated genomes and obviously the application of *in situ* hybridization techniques will be of much value here.

Virus specific proteins and antigens

Little is known about the virus specific proteins present in cells persistently infected with measles virus although fluorescent antibody methods have shown that the principal structural components may be present in considerable quantity. It would appear that sufficient proteins are available in some persistently infected cells to permit maturation of infectious virus, even in the presence of drugs which inhibit both RNA and protein synthesis. Nonetheless, the distribution of virus antigen in cells persistently infected by measles is characteristically abnormal. In many cultures, the antigens at the cell surface decrease in amount and there is an increase in the nucleocapsid antigen content of the nucleus. As cell fusion generally occurs when persistently infected cells are added to uninfected cells, it is likely that the glycoproteins are present on the cell surface, at least to some degree. Although haemolysin (i.e. cell fusion) activity can be detected, in some cases, haemagglutinin is either of low titre, or undetectable, and even haemadsorption may not be demonstrable.

The relative amounts of other virus structural proteins and "early" proteins remain to be determined.

GENESIS OF PERSISTENT INFECTION

Rima and Martin (1976) have recently reviewed the mechanisms that may result in the establishment of persistent infections by RNA viruses other than leucoviruses. The most likely mechanism is that already demonstrated for VSV by Holland and Villarreal (1974) who showed that defective interfering particles are directly involved in the onset of persistence. Alternatively, there is some evidence that the genesis of persistence results from the selection of avirulent mutant viruses or resistant cells. We shall discuss each of these viewpoints.

Selection of suitable host cells

The establishment of persistent infection in certain types of cells, particularly cells of the CNS, are of obvious importance in such diseases as SSPE and the maintenance of persistence could be related to the absence from these cells of some host-factors involved in virus matura-tion. Some evidence exists from experiments employing co-cultivation of persistently infected cells with permissive cells, which suggests that the latter provide an important factor involved in the maturation of the virus. Knight et al. (1974) have proposed that BS-C-1 cells provide an enzyme involved in the post-translational processing of virus protein, in line with the observations of Choppin et al. (1971) and Homma (1975) that trypsin treatment activates the biological activities of glycoproteins of paramyxoviruses. Whether such post-translational processing of the glycoproteins is a factor governing the maturation of virus particles, or is more related to the increase in biological activities such as infectivity is not yet known. However, experiments designed to induce higher yields of virus either by treatment with drugs or by co-cultivation cannot distinguish between the supply of factors required for maturation by the permissive cells or the removal by the dilution, resulting from co-cul-tivation with a number of permissive cells, or regulators of maturation which may be present in the persistently infected cells. The fact that, in several systems, co-cultivation with homologous uninfected cells can lead to release of virus, indicates that the lack of a specific host-cell factor cannot be the reason for persistent infection being maintained as opposed to a lytic cycle. Furthermore, in many cases the rapid kinetics of setting up some persistent cultures in the large majority of cells, which are known to be susceptible to a lytic infection, cannot be explained by the selection of resistant cells.

Selection of avirulent mutants

A good deal of information derived from various virus systems (see reviews by Preble and Youngner, 1975; Rima and Martin, 1976) suggests that mutant viruses can be released from persistently infected cells. An increase in the amount of virus yield often results from a shift of cul-tures to a lower temperature (see p. 102). In some cases, but not in all (Rima et al., 1976), these viruses appear to be temperature sensitive. Furthermore, it has seldom been shown that these "released" viruses show a marked tendency toward establishing persistent infections.

Only three reports have mentioned the successful isolation of mutants of measles virus. Bergholz et al. (1975) have isolated mutants after treatment of wild virus with mutagenic agents and found that some of the isolates were conditional-lethal mutants which under non-permissive conditions resembled SSPE virus. Gould and Linton (1975) reported that virus released from measles/HEp₂ cells did not grow at 39·5°C but grew normally at 39°C. This mutant bred true and produced predominantly miniature plaques compared to the original wild type virus. The isolates forming miniature plaques tended to set up persistent infections. Haspel et al. (1975a) have isolated 24 genetically stable t-s mutants but none of them are reported to possess a particular tendency to establish persistent infections. In no case, therefore, has there been a clearcut demonstration that t-s or any other mutants are directly involved in the establishment of persistent infections.

The role of defective interfering particles in persistence

As mentioned above there is good evidence that DI particles are involved in the establishment of persistent infection by VSV (Huang and Baltimore, 1970; Holland and Villarreal, 1974). Although it had been known for many years that undiluted passage of measles virus leads to changes in the type of CPE from syncytia to strand-forming morphology, it has only recently been demonstrated that undilute passage of measles virus will lead to the formation of DI particles containing sub-genomic RNA (Hall et al., 1974). Recent experiments on plaque purified measles virus (Rima et al., 1977) showed that after only four UP passages there was a considerable decrease in the yield of infectious virus. Furthermore, the UP viruses gave rise to persistent infections in Vero cells affecting about 80% of the cells. Similar observations have been made by ter Meulen and Martin (1976) on the continuous undiluted passage of canine distemper virus during which there was a rapid generation of carrier cells only in alternate passages, possibly related to the variation in the amount of DI particles present. These experiments show that even after plaque purification the generation of DI particles is extremely rapid and the preparation of DI-free virus stocks, by continual terminal dilute passage, is difficult since the multiple growth cycles involved and the resulting increase in size of syncytia create a situation analogous to undilute passages, during which DI particles accumulate.

The exact role that DI particles may have in the maintenance of persistent infection is still unclear. The mechanism of autointerference by DI particles can readily be explained on a competition basis between intact genome RNA and sub-genomic RNA for a limited amount of polymerase, as suggested for the VSV system by Huang and Baltimore (1970). However, autointerference itself, although resulting in a decrease in yield of infectious virus, does not always involve the establishment of persistent infections and it is difficult to see how competition alone could maintain cultures for such protracted periods.

Stanwick and Kirk (1976) have recently suggested that in LCM/BHK cultures there may be a heterogeneous population of cells, some infected, some uninfected and the bulk producing DI particles. They suggest that a dynamic equilibrium exists amongst the population of cells present and a similar analytical study of measles persistent cultures would be of interest. If such a mechanism of control is applicable to all persistent cultures, it should always be possible to clone out "cured" cells from all the cultures, presumably at a fairly constant ratio. The opposite would hold for the integrated state (p. 108). "Cure" would be a rare event.

As mentioned earlier the maintenance of persistence appears to be related to a block in the maturation process and it seems important to attempt to understand how such a block can be related to the presence of DI particles.

Rima (in discussion) has pointed out that the apparent sequential production of virus protein is probably controlled at the primary transcription stage by variation in the affinity of binding sites, on the various genes on the 50S RNA, to the polymerase. Under conditions where the amount of 50S RNA is restricted by competition for polymerase by sub-geno micRNA, only small amounts of proteins coded by low affinity genes would be synthesized. In other words, DI particles would have a relatively greater inhibitory effect on the production of those proteins which mainly depend upon secondary transcription and are formed slowly, as these require the accumulation of 50S RNA, than they would on the production of early proteins. This hypothesis implies that persistently infected cells would contain a deficiency in the proteins concerned with virus maturation (Fig. 7.1), in particular the proteins which accumulate late in infection such as glycoproteins and the membrane protein (see p. 86). On the other hand, if the sub-genomic RNA was actively translated into protein, it could play a positive role in controlling persistence by over-producing a factor that may function as a regulator in the maturation step. These two hypotheses are open to ready examination, since the former implies that DI particles will effect a selective decrease in the relative amounts of certain proteins whereas the latter would predict a selective amplification of the regulator protein(s). The over-production of a regulatory protein, resulting from the participation of sub-genomic RNA in persistent cells, could help to explain a recent observation by ter Meulen and Martin (1976) that a CDV–Vero

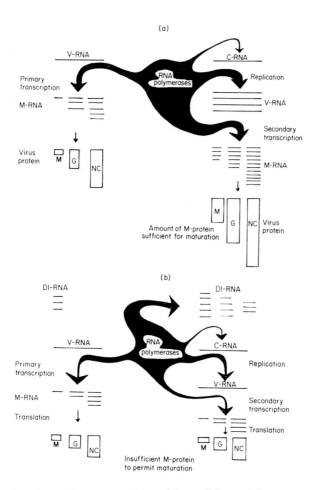

Fig. 7.1. A schematic presentation of how DI particles may maintain a persistent infection by negative control mechanism. (a) A lytic infection. The rate of maturation depends on the formation of a sufficient amount of a rate limiting protein (perhaps the membrane protein M). (b) Persistent infection. In the presence of sub-genomic RNA the production of M-protein is reduced due to a restriction of the amount of secondary transcription that occurs, when the DI RNA is competing for the polymerases.

cell line released a strong inhibitor of CDV and measles virus into the medium and that an onset of lytic crises can be initiated by removal of used medium.

Although multiple infection of cells with virion and DI particles may play a major role in setting up persistent infection *in vitro*, a problem does

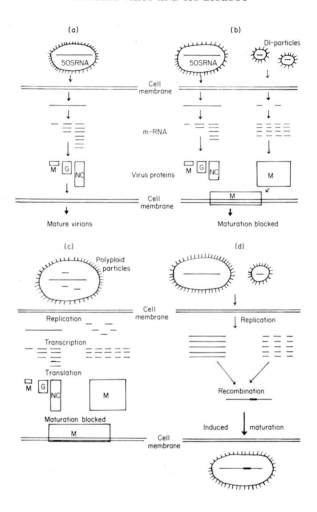

Fig. 7.2. Possible schemes of how DI particles may maintain persistent infections by positive control mechanisms. (a) A lytic cycle: optimal amounts of rate limiting factor produced. (b) DI RNA participates in protein synthesis and increases the amount of a regulating factor or an inhibitor of maturation, thus preventing the formation of virions. (c) The presence of polyploid particles, containing; both 50S and sub-genomic RNA would have a similar effect as in (b). (d) Generation of variants of measles virus by the induction of maturation in persistent cells. Recombination with specific DI-RNA could result in the amplification of genes which produce regulator or inhibitory proteins.

exist in relating this method to the establishment of persistence in the *in vivo* situation, where high primary multiplicity of infection is unlikely to occur. Since these viruses mature by budding it is probable that some particles contain multiple species of nucleocapsids, some of sub-genomic size, and, of course, such particles would have a similar biological effect to the mixture of viruses and DI material. It is possible that such polyploid particles are found more readily in certain cell types, in the CNS for example, and hence these cells could show predominantly persistent infections.

Furthermore, if DI particles are present in substantial amounts during long protracted infections, it is also possible that they participate in the generation of variants which may eventually be released or recovered from persistently infected cells or organs. Such variants could result from recombination between the intact RNA genome and sub-genomic RNA species, resulting in the release of virus containing selectively amplified genes. Of course if these genes could play a regulatory role then the characteristics of the released virus would be different, namely, they might possess a higher tendency to persistence, slower growth rates and smaller plaques. The pathogenicity of measles virus, and perhaps also minor variations in strains, may be eventually understood by the involvement of defective genomes in an active role which may help to maintain persistent states and also participate in the generation of variants which can be recovered from persistently infected cells or organs.

The various hypotheses that we have discussed regarding the role of DI particles in maintaining persistent infection, are summarized schematically in Fig. 7.2. These ideas are open to direct experimental assessment by currently available techniques and we believe that their consideration will be valuable in elucidating the detailed molecular mechanisms involved in persistent infections.

COMMENT—THE FEATURES OF PERSISTENT INFECTION

Carriage of measles virus in replicating cells is marked by certain common and interconnected features. Cytopathic effect and yield of infectious viruses are reduced or absent. In time, with continued passage, the yield of haemagglutinin also decreases, haemadsorbing cells become less numerous and haemadsorption itself less intense; fluorescent antibody reveals that virus antigen is absent from the membrane of many cells and intra-cytoplasmic aggregates or inclusion bodies become sharply demarcated within an unstained cytoplasm. Antigen, which is probably nucleocapsid, tends to accumulate in the nucleus and this is seen at its most extreme in strains of virus derived from SSPE and in SSPE itself, when the whole nucleus becomes packed with antigen. The electron microscope usually shows typical nucleocapsid material in abundance in the cytoplasm and often in the nucleus also, and non-yielding cultures have been shown to lack the apposition of nucleocapsid to the inner surface of the cell membrane which is so characteristic of the productive virus cycle. There may also be absence of budding, or budding may be noted of empty particles which also lack characteristic virus spikes.

A second general feature of the carrier state is its variability from one system to another and from time to time in the same system. An impression is given on studying these systems that a gradation in the degree of suppression of the virus growth cycle is possible, ranging from partial cytopathic effect and some virus maturation (that is, nearly complete production) to complete suppression of both these activities. There is no adequate evidence of carriage of virus genome without antigen synthesis; most measles antigen-free cells grown from carrier lines have been shown to be fully susceptible to superinfection with the same strain of measles virus. Likewise, integration of the virus genome has not been finally proven, although claimed by Zhdanov and mentioned by Simpson recently (Zhdanov and Parfanovich, 1974; Simpson and Iinuma, 1975). Both these relationships—virus suppression and integration—if they exist, have important implications for the ability of the virus-infected cell to escape immunological destruction *in vivo*.

Temperature-sensitive effects have been noted in that virus yield and cytopathic effect may increase at lower temperature and two systems have been shown to yield a more-or-less stable temperature-sensitive variant of measles virus. Reactivation of the lytic cycle occurs readily in some systems by cell fusion and to some extent by treatment with actinomycin D or other drugs.

All these manifestations point to a physiological control of the system in which production fluctuates to some extent and which is in its most stable state of suppression in a few cloned lines of infected cells and in the two systems derived from neurotropic virus. Infectious viruses can hardly be obtained from these.

Some point in the maturation process is blocked or is incomplete and it is specially significant that cell fusion studies permit the manufacture of infectious viruses from products already present in the "blocked" infected cell (Knight *et al.*, 1973). It suggests that dilution, either at the membrane or within the cytoplasm is responsible, rather than supplementation by fresh host-cell factors.

What causes initiation of the chronic carrier state has not been explained, but undoubtedly deliberate introduction of excess virus at a high multiplicity of infection permits the carrier state to become established. This indeed could be occurring naturally in surviving cells selected from low multiplicity of infection, for the last cells to be infected by natural spread of the virus would be subjected to an increasingly large multiplicity of infection by newly-produced viruses.

On the other hand, it is not easy to explain how virus-specific antibody and perhaps other agents affecting maturing surface virus, exert a stabilizing influence on the regulated carrier state unless indirectly by inhibiting cytolysis. The suppression of the cytocidal effect of measles virus is obviously important and nothing is known about its qualitative or quantitative control. One mechanism may be neutralization and removal of virus from the cell surface, recently described by Joseph and Oldstone (1975).

Neuropathogenicity of measles virus

INTERPRETATION: (1) *Laboratory strains of measles virus can cause encephalitis in young rodents when inoculated intra-cerebrally.* (2) *Decreasing susceptibility of the host with increasing age is a marked phenomenon.* (3) *Adaptation of the virus results in increasing virulence and the ability to affect adult animals.* (4) *Pathogenicity seems to depend on virus growth and extent of infection rather than localization in specific vital centres.* (5) *Estimates of virus growth are complicated by the production of non-infectious virus and by the existence in brain of non-specific inhibitors of infectivity.* (6) *Much of the spread of virus in brain occurs by cell-to-cell transfer, probably with some fusion between cell membranes.* (7) *The immune state of the host strongly affects progress of the infection and acquired immunity is beneficial. There is no evidence of immune pathogenic effects, but demyelination has not been considered.* (8) *There is preliminary evidence that infection of mature brain cells with measles virus results in abortive infection.* (9) *Measles virus which has been passed in brain may lose the ability to produce infectious virus in tissue culture until re-adapted.*

Encephalitis is the least frequent of the less rare complications of measles, much less common than bronchitis and otitis media (Miller, 1964). Nevertheless there is indirect evidence of cerebral disorder during measles, more than half of the electro-encephalographic traces being abnormal (Weber *et al.*, 1969). Amongst childhood exanthemata similar abnormalities are found in about half of the patients with measles, one-third of those with mumps, one-fifth of those with varicella and one-tenth of those with german measles (Gibbs *et al.*, 1959). The exact pathogenesis of measles encephalitis in man remains unknown. It is an acute disease, often followed by permanent and serious functional defects (much more so than mumps encephalitis), and since about 1966 subacute sclerosing panencephalitis has been recognized as a chronic infection by measles virus (Connolly *et al.*, 1967; Freeman *et al.*, 1967; Reviews: Connolly, 1972; ter Meulen *et al.*, 1972b). The two features of neurotropism and chronicity of infection have formed the basis of much research into the behaviour of measles virus before and since that date.

There is a well-recognized form of severe encephalitis, accompanied by demyelination, which is known to occur after certain virus infections including measles and apparently after vaccination against variola and rabies. Signs of acute virus invasion of neurones and of acute inflammation are absent and virus, whether measles, vaccinia or rabies, is scarcely ever recovered from affected areas of the brain. Although it also has been called "para-infectious" or "post-infectious" encephalitis, it is not known whether local infection by virus precedes its onset or not. It is not usually preceded by a meningitic or an encephalitic phase of the generalized infection. Neither the acute or chronic encephalitis considered in this section refers to post-infectious demyelinating encephalitis except to recognize that it could sometimes be a sequel to either inflammation.

EXPERIMENTAL ENCEPHALITIS

The encephalitic properties of measles virus have been studied in two different ways. The earlier way entails various endeavours to adapt the virus to serial growth in brain, mostly of mouse, rat or hamster; and the second uses encephalitis as a marker of pathogenesis or virulence for different isolates of measles virus.

Adaptation to intra-cerebral passage

Growth in mouse brain was accomplished soon after measles virus had been isolated and has been achieved in different laboratories (Imagawa and Adams, 1958; Carlström, 1958; Waksman *et al.*, 1962; Matumoto *et al.*, 1964), and there are claims to have recovered virus directly from mouse brain inoculated with patient's material (Arakawa, 1948, 1949; Greenham *et al.*, 1974). In the paper of Greenham *et al.* intra-cerebral inocula contained whole cells from biopsy specimens of infected human brain.

The Imagawa paper is interesting in that, even when growth was established in suckling mice from a HeLa-grown strain of Edmonston measles, the titre of cell-free virus, tested in mice, did not rise above $10^{1·5}$ to $10^{3·1}$ TCD_{50} and, although the cytopathic capacity against HeLa cells was retained, the 20th intra-cerebral passage, when grown on HeLa cells, released no infectious progeny virus into the supernatant culture fluid. This is probably the first evidence that neurotropic strains of the virus tend to become non-releasing. Adaptation was accompanied by a shortening of the incubation period of the disease, a rising mortality and, sometimes, gradually increasing susceptibility of adult animals. Thus Waksman *et al.* (1962) and Burnstein *et al.* (1964) have recorded in young hamsters inoculated with the Philadelphia strain of virus, a 9-day incubation period at the 6th intra-cerebral pass, 5 days at the 20th and only 2 days at the 79th. From the 65th pass onwards, adult hamsters were susceptible to the virus and so were mice. This adapted virus could cause encephalitis by the sub-cutaneous route, the more so when assisted by hyaluronidase in the inoculum.

A third adaptation to mice of yet another strain of measles virus, Sugiyama, which was originally isolated on monkey kidney tissue culture, seemed to be facilitated after passage of the virus on human amnion cells. The same requirement was found by these workers to assist the adaptation of Edmonston virus (Matumoto *et al.*, 1964). The variability of measles virus is again revealed by the fact that Matumoto's adapted measles virus affected only infant mice, less than 5 days of age; Waksman's affected adult animals also and, unlike Imagawa's and Adams' virus, grew in tissue culture of dog kidney. Another early report of successful passage in mouse-brain, which is less detailed, used a Swiss strain of measles virus, grown in human embryo kidney. This strain required 15 blind passes for adaptation, yielding virus that gave fairly low titres, about 10^2 LD_{50} by intra-cerebral assay (Carlström, 1958). The pathogenicity to tissue culture could not be tested then.

Later, Burnstein and Byington (1968) found that their rodent-adapted virus grew in suckling rat brain from which the virus could be passed to the BS-C-1 line of African green monkey renal cells only by inoculating intact, infected brain cells, whereas the homogenate was infectious on intra-cerebral inoculation into hamsters. This failure to grow may have been due to an inhibitor found in rat brain which affected the surface of cells exposed to it, in company with the virus.

Adaptation to brain does not seem to follow a common course in all species. In young rats, the rodent-adapted virus, which exhibited a sharply decreasing mortality with increasing age, did not give rise to chronic infection (Byington and Burnstein, 1973) whereas hamsters, infected with an adapted strain of measles virus that originated from subacute sclerosing encephalitis produced a few animals that developed chronic hyperactivity and myoclonus from 45 to 120 days after infection. A hamster-adapted SSPE measles virus could be recovered from the brain of some animals which showed, although rarely, inclusion body encephalitis (Byington and Johnson, 1972). The pathogenicity of two different mouse-adapted viruses has been compared in newborn mice and weanling mice. In general, no chronic symptoms developed and survivors of encephalitis gradually lost virus from the brain (Griffin *et al.*, 1974).

Chronic infection in animal brain

Subacute and chronic infections have been recorded in animal brain in addition to the acute lethal infections produced by adapted measles virus. Schwarz measles virus, grown in BS-C-1 cells, proved fatal to newborn hamsters, on intra-cerebral inoculation, but in those born to immunized dams, and thus well supplied with natural antibody, the incidence of encephalitis was one-sixth of that in unprotected animals. Virus persisted in the brains of many of the survivors. Symptoms of chronic encephalitis appeared spontaneously or could be induced by cyclophosphamide treatment of the infected animals (Wear and Rapp, 1971).

In similar experiments Katow *et al.* (1973) describe how one immunized hamster out of a group inoculated with an SSPE strain of measles virus (Niigita) developed a subacute meningoencephalitis. The virus was otherwise uniformly fatal to infant and adult unprotected hamsters.

Neuropathogenicity of non-adapted strains

The ability of different laboratory strains of measles virus to produce symptoms after being inoculated into the brain of newborn hamsters varied from total pathogenicity for brain-adapted viruses to complete attenuation of the Schwarz strain of measles virus grown on chick embryos. Yet the Schwarz virus became quite virulent (68% to 86%) when passed twice on Vero (monkey) cells. Strains recovered from patients with subacute sclerosing panencephalitis are not necessarily neurovirulent, for two were found to be poorly pathogenic to hamsters (Albrecht and Schumacher, 1971). Workers in the same laboratory also tested three strains of measles virus by intra-thalamic inoculation into monkeys that had not been exposed to natural infection; only hamster brain-adapted virus gave symptoms, and then only when the animals were treated with cyclophosphamide. Edmonston virus and its attenuated Schwarz strain produced no effect in the same species but were recoverable from the animals five months later (Albrecht et al., 1972).

A difference in neurotropism between the Edmonston and the Schwarz strains of measles virus was observed in Syrian hamsters but adaptation to 100% lethality was rapid for each strain (Janda et al., 1971). In this study, in which uniformity amongst the strains tested was assisted by passing them all first in Vero cells, adaptation to hamster brain was rapid, except when the virus used had been prepared from undiluted seed (Oddo et al., 1961). The pathogenicity of virus derived from concentrated inoculum was originally greater than that of the same strain grown from highly diluted seed, but the incubation rate was slow to decrease as compared with the rapid decrease which characterized adaptation of the latter type of virus. Several non-neurotropic measles virus preparations with different cultural histories were very lethal to newborn Syrian hamsters in the experiments of Katow and his colleagues, although there were big differences in the amounts of virus recoverable from brain; some non-lethal strains increased in titre intra-cerebrally at the third day after inoculation; one SSPE strain (Niigita) was lethal to hamsters and highly virulent to mice after 10 passes in co-culture with Vero cells (Katow et al., 1973).

Hamsters were also used to compare the neurovirulence of strains of measles from subacute sclerosing encephalitis patients with the virullence of laboratory viruses. On this occasion, unlike Albrecht and Schumacher's strains, the pathogenicity of the two SSPE strains differed, neither being totally lethal and both producing more encephalitis than the laboratory viruses Edmonston and Woodfolk which had been prepared in CV-1 cells (Lehrich et al., 1970). In Shishido's experiments on the neurovirulence of measles virus in mice, the lesser susceptibility of that animal to laboratory strains of virus, as compared with the hamster, seems to be confirmed (Shishido et al., 1973). Young dogs, injected with wild and human encephalitic SSPE strains of virus, showed signs of encephalitis in 10 to 30 days following inoculation only of SSPE viruses; the incubation period was 2 to 6 months in ferrets (Notermans et al., 1973).

There is little to be learnt from these numerous observations from which only two conclusions are possible. Either different isolates of measles virus represent wild viruses possessing different degrees of neuropathogenicity, or the passage history has produced the variety of virulence that has been described. Nothing has been done to decide between these possibilities, but the latter seems more likely, to judge from the variety of effects shown by derivatives of the original Edmonston strain. The experience of Japanese workers showing that passage on their line of FL (human amnion) cells enhanced the adaptability of the measles strains to grow in mouse brain does not seem to have been examined further, but supports the suggestion that the nature or the number of sub-cultures may alter the degree of neuropathogenicity.

THE EFFECT OF AGE OF THE HOST ON SUSCEPTIBILITY

Many of the works described above give data on the effect of age of the host on resistance to intra-cerebral inoculation.

Early reports on the susceptibility of suckling mice mention increasing mortality without rise of titre, as passage was continued (Imagawa and Adams, 1958) and Waksman's report implies that full adaptation of hamster-neurotropic measles virus accounts for its pathogenicity for adult mice also (Waksman et al., 1962). In 1964 Matumoto showed that suckling mice developed a complete alteration of susceptibility to a Japanese strain of measles virus between the ages of 5 days, when they were fully susceptible, and 7 days, when they were fully resistant. The inoculum of virus was of high titre, 10^5 or more MLD_{50} (Matumoto et al., 1964).

Further evidence on age and susceptibility comes from viruses adapted to, or undergoing adaptation to, hamsters and mice. It seems that hamster-neurotropism and mouse-neurotropism usually co-exist in a virus strain, but need not do so. For example, the Schwarz strain of Edmonston virus and the Edmonston strain, grown on human amnion cells were each found to be highly virulent towards hamsters, but not to newborn albino mice (Shishido et al., 1973). On the other hand a comparison of two strains, hamster-adapted neurotropic Philadelphia (HNT) virus and a mouse-adapted neurotropic Edmonston virus, showed that the HNT virus was more lethal to infant (100%) and weanling mice (31%) than the Edmonston (93% and 22% respectively). The intra-cerebral doses of the adapted viruses were carefully adjusted (see Table 8.1) to allow for the greater infectious titre of HNT viruses (Griffin et al., 1974).

These strains and a few others, had been tested on hamsters and rats of different ages, and, as with mice, the abrupt change of susceptibility with age found by Japanese workers is not characteristic.

A pronounced, but also gradual, decrease in susceptibility with age has been seen in hamsters, although no detailed work is available. Little effect was seen on 28-day old hamsters, and no effect in 42-day old hamsters, when Edmonston virus was being adapted to intra-cerebral growth; it was quite virulent to infant animals (Janda et al., 1971, Table 8.1).

Adult hamsters gave no response to a hamster-adapted virus recovered from a subacute sclerosing encephalitis patient (HBS) (Byington and Johnson, 1972) although a non-yielder strain of similar origin (Niigita) passed fifteen times by co-cultivation with Vero cells, proved lethal to all newborn and adult hamsters and also to most newborn and adult mice (Katow *et al.*, 1973). In the experiments of Byington and Johnson, reduced pathogenicity as compared with the Japanese strain, was not due to failure to replicate, for increase of titre intra-cerebrally was demonstrated between the second and sixth day after inoculation.

TABLE 8.1

The susceptibility of rodents at different ages to intra-cerebral measles virus

A. RATS (Item 17, Table 8.2 refers).

Age	No. of rats	% affected by virus
2D	33	100
7D	23	91
14D	44	48
21D	14	0

B. HAMSTERS (Item 8, Table 8.2 refers).

Age	% affected by virus strain	
	Edmonston	Schwarz
7D	100	53
14D	50	33
28D	16	0
42D	0	0

More precise work with the hamster neurotropic derivative of Philadelphia measles virus in rats revealed a gradually increasing resistance to intra-cerebral inoculation from the second to the fourteenth day of age, when the incidence of symptoms was halved and by the twenty-first day of age rats were resistant (Byington and Burnstein, 1973; see Table 8.1).

PATHOGENESIS OF ENCEPHALITIS

It is conceivable that the various degrees of neuropathogenicity produced by separate strains of virus result from selective distribution of individual virus strains into areas of the brain which have different importances for survival of the host. Another possible explanation is variation in the total amount of virus produced in the brain. Restriction of spread and a poor yield of virus might reasonably be expected to characterize less virulent strains; ability to disseminate widely or to infect vital centres might also typify virulent strains.

Distribution of virus and lesions in brain

One way of testing this is to examine lesions in encephalitis of decreasing severity, produced by neurotropic viruses which exhibit a marked tendency to be less virulent in older animals. Both limited spread and selective localization were indeed found quite soon after adaptation to rodents was tried, for it was noticed that in older hamsters and mice the neurotropic virus was localized to certain areas, such as the hippocampal gyrus, but choroidal epithelium was spared from invasion; in very young animals virus was widespread in the brain (Waksman et al., 1962). This sort of limited spread in older animals was described also by Griffin et al. (1974), using two strains of virus and comparing suckling and weanling mice. The absence of virus antigen from the choroid plexus of 2-day old hamsters was again noted by Baringer and Griffith (1970). Thus selectivity and extent of dissemination are separate attributes of neuropathogenicity.

Virus can be located by the specific nature of measles inclusions and by immunofluorescent tracing, both of which have shown it to be capable of spreading widely in the brain following intra-cerebral inoculation. All parts do not show similar lesions. For example, Waksman et al. (1962) noted that in hamsters, giant cells and typical measles inclusions were readily seen in the ependyma, but not in glial tissue, although glial cells and neurones were both killed by the infection. No giant cells were found in mice, otherwise the spread of virus was similar to that in hamsters.

In general, the cortex and hippocampus were infected in both hamsters and mice when there were signs of encephalitis (Baringer and Griffith, 1970; Janda et al., 1971; Griffin et al., 1974). Infection of the cortex and cerebral ganglia occurred a day or two before the hippocampus and cerebellum became positive to the fluorescent antibody test (Janda et al., 1971), but there was a tendency towards sparing of the cerebellum and brain stem except when animals were very young or the strain of virus was very virulent (Matumoto et al., 1964; Byington and Burnstein, 1973; Waksman et al., 1962). The species of animals used in these tests is shown in Table 8.2. Symptomless encephalitis exists and a distribution of lesions and antigen similar to that in lethal encephalitis was found by Matumoto et al. (1964), so that regional effects alone do not account for all pathogenicity. This was also noticed by Albrecht and Schumacher (1971). Only Mirchamsy has claimed that the histological changes produced differ sufficiently enough between adapted and non-adapted strains

TABLE 8.2.

Data on experimental neurotropism

Reference	Virus strain[a]	Dose or virus titre	Host	Lesions or Site[a]	Presence of antigen or live virus
1. Waksman et al. (1962).	Philadelphia 1st isolate 2 × MK	?	Hamster, suckling mice	In suckling: Widespread in ependyma, nerve cells, glia. Giant cells conspicuous in ependyma, not in glia. Inclusion bodies in nucleus and cytoplasm. Infiltration and perivascular cuffing. In older animals: Restricted. Hippocampus and nuclei in 3rd verticle infected. Choroid plexus spared. Mice: As hamsters, but no giant cells.	Not done
2. Matumoto et al., (1964.)	Sugiyama Tanabe } × MK Tanaka Edmonston	10^2–$10^{2.5}$ $10^{2.5}$ Not tested $10^{2.8}$	Mice, suckling	No localization. Giant cells and cell degeneration throughout brain	Live virus in brain, not in other organs. Round needle-track D1. Contralateral hemisphere D4–D5. At illness-mantle-cortex, nuclei in brain stem. Ammon's horn always, even in non-sympto-matic mice. Cerebellum, medulla cord:—NOT infected. All antigen cytoplasmic. In nerve cells, sometimes glia.

		Dose	Animal	Pathology	Notes
3. Wear et al. (1968)	Imagawa's and Adams' 1958 strain	$A = 6.8 \times 10^4$ pfu or $B = 6.8 \times 10^2$ pfu	Mice, newborn	A. Degenerate cells—needle track, D3. Eosinophilic cytoplasm, horse-shoe nuclei D4. Both hemispheres, focal and general D5. Cerebellum, brain stem, olfactory bulb—not affected. B. As group A but 2 days later. Plus lesions in olfactory bulb. No inclusion bodies. No syncytia	By E. M. 110–115 A° diameter tubules in cytoplasm. Nucleo-capsid, by negative stain. No mature virions seen.
4. Baringer and Griffith (1970)	Edmonston × 2 human embryo kidney	$10^{5.5}$ TCD50/0·1 ml	Hamster newborn	Homolateral cortex D3 Contralateral D4. Cortex and hippocampus D5. Necrosis in Ammon's horn D6. Giant cells ependyma and subependyma. Correlation with giant cells D7.	Contralateral hemisphere-: In ependymal and subependymal giant cells D2. Deep spread and meninges D3. In dendrites. Not in choroid plexus. EM—virus particles in neurones and dendrites.
5. Lehrich et al. (1970)	SSPE—LEC "—JAC	$10^{2.3}$–$10^{3.3}$ pfu $10^{2.6}$ or 3×10^4 cells	Hamster 24–72 hr.	Perivascular cuffing. Sometimes confined to hippo-campus. Sometimes death without lesions.	Seen in one hamster. No details.
	Measles, wild Measles, Edmonston	$10^{2.6}$ pfu	Hamster adult		

TABLE 8.2 (*continued*)

Reference	Virus strain	Dose or virus titre	Host	Lesions or Site	Presence of antigen or live virus
6. Wear and Rapp, (1971)	Schwarz Edmonston Imagawa strain.	5×10^4 to $2\cdot5 \times 10^6$. $7\cdot5 \times 10^{1.0}$ to $8\cdot7 \times 10^5$.	Hamster newborn. Mice, newborn.	Not done.	Multiplication intra-cerebrally in both species.
7. Albrecht and Schumacher (1971)	Several Edmonston-derived strains 2 SSPE strains	?	Hamster 3–4D. Mouse, 1D.	Grade (1) Oedema, white matter disorganized. (2) Neuronal and glial degeneration in cortex, some vascular cuffing. (3) As grade (2) but spread with no predilection for any area, but cerebellum and stem less affected. Cytoplasmic inclusion bodies in some neurones. (4) Necrotizing, localized, little inflammatory reaction.	HNT (hamster neurotropic) Shows true neurotropism neuronal and glial. Meninges, ependyma and choroid NOT stained.

Reference	Virus	Dose	Animal		
8. Janda et al. (1971)	Edmonston Schwarz Oddo UP „ DP (all Vero-grown).	?	Hamster 7D–42D.	Ependyma, then deeper in cortex cell necrosis seen. Hippocampus. Giant cells. DP virus tend to go deeper into cortex.	Antigen cytoplasmic and in dendrites of neurones. Ependyma. D1. Meninges and longitudinal fissure D2. Sub-ependyma, hippocampus D3. Note radially oriented "stripes" of infected cells.
9. Johnson and Byington (1971)	SSPE Hamster adapted	$2 \times 10^{4.4}$ TCID50	Hamster newborn.	Giant cells. Ependyma, sub-ependyma D3. Inflammatory infiltrate and giant cells in cortex D4. Cells with intra-nuclear inclusions near giant cells.	Ependyma and sub-ependyma D1. Cytoplasmic antigen, not nuclear, predominantly in neurones. Spread to cortex by connecting strands of fluorescent cells.
10. Albrecht et al. (1972)	Edmonston Schwarz	about $10^{5.0}$–$10^{5.6}$	Monkey (Macaca).	With cyclo-phosphamide: Encephalitis.	HNT much antigen seen Edmonston (none).
	HNT (Philadelphia)	pfu intra-thalamic	cyclo-phospha-mide-treated.	HNT. Moderate histological inflammatory change. Edmonston—no effect. Schwarz—no effect. Without cyclo-phosphamide: Much inflammation. No encephalitic symptoms.	Schwarz (none) but virus recovered from brain 5 months later. Very little antigen seen.

TABLE 8.2 (continued)

Reference	Virus strain	Dose or virus titre	Host	Lesions or Site	Presence of antigen or live virus
11. Byington and Johnson (1972)	SSPE (HBS) Edmonston	$10^{5.4}$ TCID50 per 0·1 ml. Inoculum = 10^3. TCID50 mostly	Hamster, weanling 2 × intra cerebral 5 × BS-C-1	SSPE Acute. Intense inflammation 4D-6D. Declines after 12D. 21D, OR Chronic—Foci of astrocytes and plasma cells. 12D. Intra-nuclear inclusion bodies. 12 wks old. Edmonston—No effect.	SSPE Acute and chronic: Ependyma and periventricular cortex. Occasionally in hippocampus. Antigen in cytoplasm and nucleus of cells. Edmonston—Not seen.
12. Mirchamsy et al. (1972)	4 commercial vaccines	0·02 ml of $10^{3.5}$ TCD 50/ml.	Hamster suckling	Degenerative processes with giant cells. Different pattern from brain-adapted virus. No great detail.	Not done.
13. Schumacher et al. (1972a)	HNT (Philadelphia) 109 × IC pass suspension	0·03 ml. brain suspension	Rat, 4–11 wks. A. non-immune, B. hyper-sensitized C. and suppressed hyper-sensitized.	A. Encephalitis of white matter D19 to D21, end of expt. B. Some encephalitis of grey matter D19. Disappeared by D31. C. Slight encephalitis D19 proceeding to condition of group A at D31.	Not done.

14. Byington and Burnstein (1973)	HNT Philadelphia × 109 intracerebral	?	Rat, suckling.	Marked meningeal reaction over areas of focal necrosis D3. Massive necrosis cerebrum and olfactory tract D7.	100% encephalitis Antigen in outer layer of cerebrum D2. Maximum virus titre D7.
15. Katow et al. (1973)	SSPE Niigita	No cell free virus	Mouse, newborn Mouse, 5 wks. Hamster, newborn: Hamster, 7 wks.	Neuronal damage. Parenchymal and perivascular cell-infiltration. Paraventricular cortex and gyrus dentatus.	Antigen in cytoplasm. Mostly neurones and processes.
16. Shishido et al. (1973)	9 strains of which 4 vaccine, 1 "wild" virus in monkey lymph node does not grow in Vero.	?	Hamster, newborn. Hamster, suckling, 3–15D. Mice, newborn.	Parenchymal infiltration of cells in cerebrum. Perivascular cuffing. Giant cells in thalamus and grey matter.	Cytoplasmic staining exclusively. Seen in bodies of neurones and dendrites.

TABLE 8.2 (continued)

Reference	Virus strain	Dose or virus titre	Host	Lesions or Site	Presence of antigen or live virus
17. Griffin et al. (1974)	Mouse-adapted Edmonston. Philadelphia (HNT)	10^3 MLD50. $10^{2.1}$ MLD50.	Mice, 10–12D. weanling 4 wks.	Suckling: Meningeal and perivascular inflammation, polymorphonuclear cells. Giant cells in neocortex and hippocampus. Weanlings: No, or very little, infiltration with cells. No demyelination.	Edmonston—in suckling mice Cerebral cortex, basal ganglia and hippocampus D6–D8. In weanling In spleen 6D–18D In brain 13D. HNT—in suckling mice Foci 3D in cortex and basal ganglia. Rapid spread to hippocampus and cerebellum. In weanling—spleen D3–D15 brain D5 Cytoplasm of cells in cerebral cortex, basal ganglia and hippocampus.

^{a}MK = Passed in monkey kidney culture; D = Day

of viruses to be characteristic and perhaps useful for the study of post-infectious encephalitis (Mirchamsy *et al.*, 1972). Rozina (1971) has put forward a combined histological and anatomical set of criteria for the same purpose, and Buynak *et al.* (1962) emphasized the ability to disseminate as a criterion of virulence in monkey brain.

Measles virus certainly damages neurones (Matumoto *et al.*, 1964; Baringer and Griffith, 1970; Albrecht and Schumacher, 1971; Shishido *et al.*, 1973; Katow *et al.*, 1973), and virus and virus antigen have been seen in both the body and in the processes of the cell (Baringer and Griffith, 1970; Janda *et al.*, 1971) although Matumoto *et al.* (1964) did not find antigen in cell nuclei with their mouse-adapted virus. Glial cells are similarly affected (Albrecht and Schumacher, 1971). A strain (HBS) derived from a patient with subacute sclerosing encephalitis produced antigen in the nucleus of ependymal, cortical and hippocampal cells when inoculated into hamsters (Byington and Johnson, 1972), and the infections often became chronic in weanling hamsters.

The mode of spread

So far as the spread of virus can be traced by fluorescent antibody, it seems to affect the ependyma first, after intra-cranial inoculation, and spreads deeper from that site as the disease progresses (Baringer and Griffith, 1970; Janda *et al.*, 1971). The known occurrence of ventricular dilatation during encephalitis (Albrecht and Schumacher, 1971) and a report that some strains may damage ventricular ependyma preferentially (Haspel and Rapp, 1975) suggests that virulence could depend occasionally on "tropism" for different regions of the brain. Radially arrayed "stripes" of fluorescent tissue, seen by Janda *et al.* (1971), suggests that passage along cell tracts occurs, presumably by cell-to-cell transfer, and virus may go deep into the cortex. It also spreads to the uninoculated hemisphere of the brain (Matumoto *et al.*, 1964; Baringer and Griffith, 1970).

Spread of virus from one part to another may be slow in adult animals. Even in suckling rats measles virus antigen has been seen in cells of the superficial layers of the cerebral cortex 2 days after infection and as many as 5 to 8 days before the death of the rat from widespread extension of the virus. According to Matumoto *et al.* (1964) cells of Ammon's horn in mice were invariably infected, even during symptomless encephalitis. The use of such a constant target suggests itself as one means to study the mode of spread of measles in the brain, especially to decide the importance of cell-to-cell, as opposed to extracellular, transmission. The effect of increasing age in reducing the extent to which virus spreads offers similar opportunities, for adult nerve cells are known to produce long-lasting, non-lytic productive cycles after infection with any of several enveloped viruses. There is, therefore, less chance of flooding interstitial spaces with large quantities of infectious viruses, and tracing of antigen from cell to cell becomes significant.

Few attempts have been made to follow virus in other organs of the body after intra-cerebral inoculation, but Griffin and his colleagues (1974) did note that two strains of neurotropic virus of different pathogenicity inoculated intra-cranially, could be traced in the spleen of weanling mice a few days before it could be found in the brain. They also noted that the cellular reactivity in weanling mice was inconspicuous, and that immuno-suppressive measures did not increase the death rate.

Local cellular reactions

During the encephalitic process inflammation is present, judging by infiltration of tissue with leukocytes and by perivascular "cuffing" with white cells. Given that symptomless infection does occur (Matumoto et al. 1964; Albrecht et al., 1972), it is interesting to note that, when symptoms clear up, inflammation often resolves (Byington and Johnson, 1972). In special circumstances, when monkeys have been made immunologically inactive by means of cyclophosphamide, inflammation can be shown to be absent, and the symptoms of the disease well marked; in immunologically non-suppressed monkeys, inflammation is excessive and signs of encephalitis are absent (Albrecht et al., 1972). Presumably the inflammatory process acts to cure, not to produce the disease. Other experiments in immunized adult rats (Schumacher et al., 1972a), which normally get a non-lethal subacute infection by inoculation with hamster-neurotropic measles virus, have shown that immune mechanisms associated with hypersensitivity end the pathological progress before 31 days, as compared with the longer lasting process in normal rats. Rats whose hypersensitivity was temporarily put out of action by immuno-suppressive drugs suffered the same clinical and pathological processes as non-hypersensitive animals.

It is possible to gain some impression of factors common to measles encephalitis by consulting the summary of data from papers to which we have referred (see Table 8.2). Some of these did contain information about the amount of virus replication intra-cerebrally which we have omitted because of uncertainty about its significance. Reports of abortive infection in brain cells and of the presence in brain of inhibitors of virus infectivity mean that growth curves will have to be re-assessed in conjunction with these new facts in every experiment.

The information in the table does suggest that increased virulence is accompanied by wide-spread infection and, presumably, a greater total synthesis of virus antigen, rather than by localization to specific areas of the brain. Such areas as are commonly selected by measles virus in mice and hamsters, e.g. the hippocampus or Ammon's horn, are very similar to these selected by other neurotropic viruses, such as neurotropic influenza virus. On the other hand the sparing of certain areas, the cerebellum in older mice and the choroid plexus at all ages, indicates that certain specific changes do occur with developing maturity of the brain. It also makes it quite unlikely that the pathogenesis of measles encephalitis is secondary to changes in blood vessels and the blood-brain barrier as it seems to be in lymphocytic chorio-meningitis.

There is a hint that SSPE strains of measles virus infect neurones more readily than do standard strains of measles virus, but the time factor influences this appreciation as the evidence applies mostly to subacute infection which neuro-adapted wild strains of the virus rarely produce. The reports that SSPE strains do grow in the nucleus of neurones are convincing. Fluorescent antibody studies favour cell-to-cell transfer as the common mode of extension of measles infection to various regions of the parenchyma and this method not only permits the spread of incomplete virus but has been seen in subacute sclerosing panencephalitis (p. 146).

SIGNIFICANCE OF FINDINGS

Experimental encephalitis produced by intra-cerebral inoculation is an unnatural process in that it presents infection as spreading to deeper parts of the brain from the meninges and ependyma on which virus has been deposited artificially. Whilst the method of cell-to-cell spread, which experiment has illustrated, is likely to take place also in man, it is unlikely that human measles encephalitis starts with infection of meningeal cells. There are no clinical data which indicate it. Lodgement in the brain is almost certainly a viraemic process and it should be possible to obtain evidence of growth of virus through vessel walls or of the passage of carrier leukocytes through capillary walls into the tissue spaces. We could not find evidence of either process having been observed in SSPE, nor in animal experiments, although one neuro-adapted strain of measles virus has been reported as neuropathogenic after being inoculated extra-cerebrally.

One report of a mutant of measles virus which grows selectively in the ependyma, with consequent damage, means that particular tropism for specific structures in human brain cannot yet be ruled out as pathogenic factors. It has been customary to invoke immunological processes to explain the cytopathic affects of measles virus in subacute sclerosing panencephalitis, as has been done in slow virus infections such as Aleutian disease of mink or lymphocytic chorio-meningitis of mice. If anything, the publications quoted in this section suggest that measles virus is sufficient cause for the symptoms produced in acute and in chronic encephalitis. The immune response may be insufficient to eradicate the virus and so allow chronic infection to become established. Experimental suppression of the immune response seems to favour reactivation and spread of the infection in all circumstances in which it has been tried, whereas stimulation of the specific immune response, though rarely studied, has encouraged elimination of cerebral infection (Albrecht et al., 1972; Schumacher et al., 1972a). We do not extend this impression to demyelinating disease following measles in which the part played by the immune response is unknown.

Introduction of measles virus into a host, in a non-productive, cell-associated form, provokes little immune response and also avoids immunological attack. Examples of this are seen in encephalitis due to SSPE strains of measles virus (p. 155). Experiments with these strains have also provided good evidence of the correlation between advancing immunological maturity and the lessening susceptibility to neuropathogenic strains of measles virus.

That interplay of these two factors—incomplete virus replication and immunological control—determines the outcome of an infection seems to be fairly clearly indicated, but it is incorrect to suppose that positive correlation between resistance and increasing age involves immunological mechanisms alone. Indeed it has recently been shown that the neuro-adapted HBS strain of Philadelphia 26 measles virus not only fails to produce live virus in many weanling mice, but also fails to assemble any recognizable virus structures. Neither inclusion bodies, nucleocapsids or virions may be seen in cells which, however, do show evidence of abundant production of capsid antigen and antigen of the membrane-haemagglutinin complex (Herndon et al., 1975). Abortive infection of this sort, related to increasing age, has also been reported in mouse brain with influenza A, another enveloped virus (Fraser, 1959). This mechanism alone would limit spread of infection.

The existence of a measles virus-inhibitor in cerebral tissue is a third factor to be taken into account. It may make the absence of infectious virus in some experiments more apparent than real; the ability of intra-cerebral inoculation to detect measles virus produced in brain cells, when tissue cultures prove negative, has been referred to. Whether the same inhibitor has any intra-cellular influence in producing non-infectious virus or even in interfering with some earlier stage of morphogenesis is not known. It may well explain the relative rarity with which live measles has been recovered from patients with measles encephalitis.

These recent discoveries of incomplete virus growth and of virus inhibition by substances in brain re-open some interesting problems of the relationship between measles virus and human disease, particularly the question of developmental maturity and the incidence of acute and of chronic encephalitis.

9 | Measles virus and subacute sclerosing panencephalitis— SSPE-viruses

INTERPRETATION: (1) *Measles virus can be isolated from brain in subacute sclerosing panencephalitis.* (2) *The isolated strains vary in their properties as much as do wild strains of measles virus.* (3) *There is morphological evidence of arrested virus assembly.* (4) *It is not known whether arrest is imposed at the site of assembly or whether it arises from earlier defects in the growth cycle.* (5) *Cell culture and cell fusion gradually release the virus from restricted multiplication.* (6) *In experimental models of the disease, immune function has a decided influence on the outcome of the disease.* (7) *Other factors, such as type of cell infected, developmental state of the brain and influence of other infections, need to be studied.*

ASSOCIATION BETWEEN MEASLES VIRUS AND SSPE

Subacute sclerosing panencephalitis (SSPE) is a rare, chronic and usually fatal inflammation of the brain which has been known in the past by several different names such as inclusion-body encephalitis (Dawson, 1933) or subacute sclerosing leukoencephalitis (van Bogaert, 1945), and which remained a disease of unknown aetiology until about 10 years ago. Since then histological, electronmicroscopic, serological and virological studies of the brain all indicate that measles virus is constantly present in neural and glial cells and the disease may be properly thought of as a chronic measles encephalitis. However, the precise nature of the association between virus and cell in this pathogenic state is not known, nor is it strictly certain, although generally assumed, that the virus has been present in brain cells since the attack of measles which has usually preceded the nervous condition by several years.

We therefore give an account of the principal observations made during detection and recovery of measles virus strains from a number of SSPE patients. By comparing these to some *in vitro* properties of carrier cultures of measles virus, it is possible to draw some tentative conclusions about the semi-persistent state of measles virus in the pathogenesis of subacute sclerosing panencephalitis. For convenience, the lazy abbreviation SSPE viruses will be applied to measles viruses isolated from SSPE patients. (Reviews: Connolly, 1972; ter Meulen *et al.*, 1972b.)

Cell culture and recovery of virus

Detailed histological findings on cerebral tissue fit well with the cellular changes that typify both inclusion bodies of an RNA virus and the tubular structures like those made by a virus of the measles group (Herndon and Rubinstein, 1968; Perier *et al.*, 1967). Unfortunately, well-conducted experiments failed to recover virus from homogenates of human encephalitic brain (Adels *et al.*, 1968; Harter and Tellez-Nagel, 1968; Perier *et al.*, 1968), a fact that emphasized the importance of a successful, contrasting technique. It soon became apparent that cultures of brain taken from patients with subacute sclerosing panencephalitis, or inoculation of uncultured intact brain cells into cultures of animal cells of various species, gave rise in the first instance to propagable cell lines that contained either measles antigen or infectious measles virus; cytopathic effects in the form of typical syncytia were often present (Chen *et al.*, 1969; Horta-Barbosa *et al.*, 1969a,b; Payne *et al.*, 1969). Cell growth and, perhaps, cell-fusion are probably critical factors in releasing the latent virus. Homogenate of brain can be infectious, as was first shown by Katz and his colleagues who isolated an agent transmissible in ferrets (Katz *et al.*, 1968, 1970a).

These successes, which were sometimes marked by attendant difficulties in propagating the newly-recovered agents, stimulated speculation about the exact nature of the virus in spite of its strong antigenic relationship to measles virus. Even when identity between measles virus and the SSPE virus strains became accepted, the possibility remained that strains of encephalitic origin were neurotropic variants or mutants of wild measles viruses. It is not easy to refute or to establish this belief, for the longer that strains take to become established in the laboratory, the more likely they are to acquire by adaptation the characteristics of the laboratory-grown measles virus. It is quite certain that no virus derived from subacute sclerosing encephalitis has been shown to retain unique characteristics by passage at limit dilutions in the same way that the 'O' type of haemagglutination by classical influenza A virus can be maintained distinct from the chick embryo-adapted, or 'D', character. A slight lack of sensitivity to standard measles antibody may be an exception. However, differences of behaviour as compared with wild measles exist, more especially in the early stages of cultivation of new isolates.

Difficulties in isolating SSPE viruses

Given the need for co-cultivation in recovering SSPE measles viruses from infected brain, success is not automatic, even when the tissue used is known to contain antigen or morphological components of virus in abundance (Katz and Koprowski, 1973). The first apparent sign of re-activation is cell-fusion, either in cultures of infected brain cells or in co-cultivation with measles-susceptible cells, and it seems to be a relevant corollary of the need to supply a susceptible cellular substrate, either by cell-division or by co-cultivation, that released virus does not at once become available. Co-cultivation of SSPE brain with BS-C-1 monkey cells failed to yield infectious virus in the first twelve sub-cultures (Fig. 9.1; Payne et al., 1969); no infectious virus was yielded from the cultures of infected brain cells (Oyanagi et al., 1970) and cytopathic effects only appeared at the ninth and twelfth sub-cultures of two others (ter Meulen et al., 1970). All cultures contained measles virus antigen or nucleocapsids.

To determine the exact stage in a series of sub-cultures at which mature virus first appears is often difficult for, even when released from tissue culture, the progeny virus may have no great affinity for cultured brain cells, no matter whether these are derived from human or hamster brains. This failure to infect seems to be a result of failure to adsorb to the cell surface (ter Meulen et al., 1972a, 1973). Also, when, as a result of co-cultivation experiments between brain cells from subacute sclerosing encephalitis and green monkey kidney cells, cell-free virus could not be recovered in culture, it could be found by inoculating the same material intra-cerebrally into suckling hamsters (Parker et al., 1970). This lack of affinity need not be regarded as unique to virus derived from subacute sclerosing panencephalitis, for it may be acquired quickly after passage of laboratory strains of measles virus intra-cerebrally in animals and, in addition, it has been associated with inhibiting substances in brain tissue (p. 120). Its value as a genetic marker of unique SSPE variants is therefore very doubtful.

Delayed maturation

There is, however, no doubt that the maturation process is partially or wholly suspended *in vivo* and in the early steps of propagating SSPE viruses *in vitro*. The evidence is morphological.

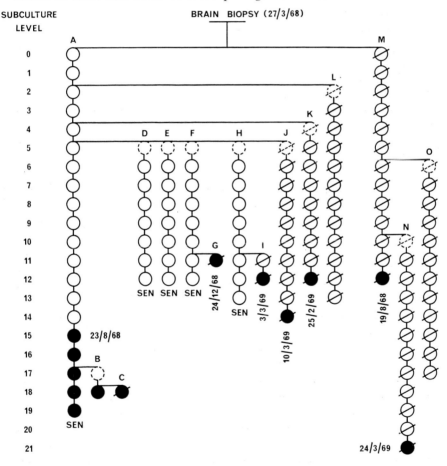

F IG. 9.1.   The chart illustrates well the considerable and variable number of passes or sub-cultures which may be required before surface antigen (● = specific haemadsorption) is expressed by SSPE virus. (From Payne *et al.*, 1969, with permission).

The normal maturation of measles virus, is marked by thickening of the plasma membrane of the infected cell, by a distinct alignment of nucleocapsid under the thickened membrane, which also differentiates into the fringed structures which resemble the spikes of the virion, and by budding of these special areas into processes which are pinched off and released as virions (p. 93, Fig. 6.8). The same morphological changes are characteristic of the growth of measles virus in cells cultured from the nervous system or *in vivo* (Baringer and Griffith, 1970). Growth of the Edmonston strain of measles virus has been studied in cultures of hamster cerebellum and in cultures of spinal dorsal root ganglia of the hamster and displayed the normal process of virion formation by budding

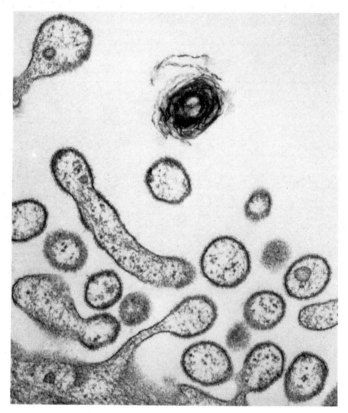

FIG. 9.2.   Virions in extracellular space of organculture of hamster cerebellum, 3 days after infection with SSPE virus. Membranes have differentiated into dense inner margin and blurred outer coat, but nucleocapsids are many—fewer than seen after infection by measles virus. (From Raine *et al.*, 1973, with permission).

of the cell membrane and enclosure of nucleocapsids. These are long-lasting cultures. The only difference noticed between them was a tendency for the cerebellar culture to produce particles during the late stages of growth which were empty of nucleocapsids, a trend which was not apparent in subsequent experiments (Raine *et al.*, 1969 1971).

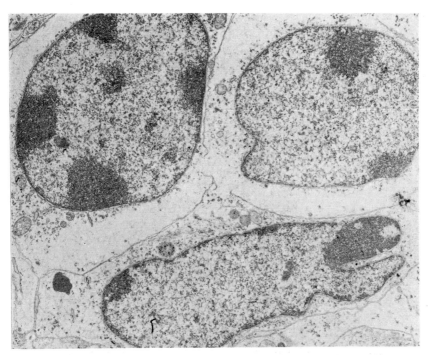

FIG. 9.3.   Section showing discontinuity of the plasmalemma between two granule cells and one astrocyte-like cell in organculture of hamster cerebellum infected with SSPE virus. It shows that the membranes are doubled back, as in fusion, not broken. (From Raine *et al.*, 1973, with permission).

The morphogenesis of virus in brain cells, bearing virus and cultured from SSPE material, was examined by Oyanagi and his colleagues. One SSPE strain, Rob, grew in cells which showed a cytopathic effect and which contained bundles of nucleocapsids, but budding was absent and no virions were seen (Oyanagi *et al.*, 1970). A comparison between two other SSPE strains, JAC and LEC (Table 2.1), and standard measles viruses, all grown in CV-1 monkey kidney cells, showed that although the SSPE strains produced plenty of rough tubular filaments, diameter 22 nm, in the cytoplasm of infected cells, they were never aligned along the under surface of the cell membrane. Furthermore, there were no fringed structures like virus spikes on the membrane although there was some budding of empty particles. In the nucleus of cells there were aggregates of material forming inclusion bodies. By comparison the standard viruses, one Edmonston and one freshly isolated strain of measles virus, formed no intranuclear inclusions, although smooth tubular filaments were seen there. There was normal budding of spiked particles containing capsids, which were aligned under the envelope of the virion. Fluorescent antibody was used to confirm that the intranuclear inclusions formed by the JAC and LEC strains of virus contained measles virus-specific antigen (Oyanagi *et al.*, 1971).

Further support for the occurrence of an abnormal maturation process during infection by SSPE virus was added by studying the morphogenesis of the Mantooth SSPE virus in cultures of hamster cerebellum and *in vivo*. This virus produced nucleocapsids, the cell membrane became thickened, but no alignment of nucleocapsids took place and budding of

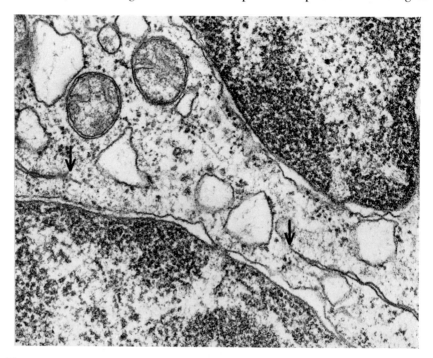

Fig. 9.4. Between two cells, from a human brain biopsy, short arrows point to two opposing plasma membranes. The absence of continuity between the left and right pairs of membranes indicates cell fusion. (From Iwasaki and Koprowski, 1974; with permission).

the thickened membrane produced only empty virions (Fig. 9.3) (Raine *et al.*, 1974a). The same process was seen when chronic infection was studied in the brain of hamsters which had recovered from acute encephalitis produced by the Mantooth virus (Raine *et al.*, 1974b). Nevertheless spread of SSPE virus is possible by cell fusion which has been shown to occur (Raine *et al.*, 1973; Fig. 9.4).

The failed maturation which the foregoing papers report experimentally, has its counterpart in man. One careful microscopic study

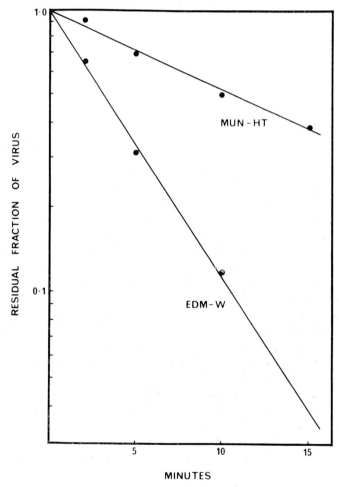

FIG. 9.5.   Kinetics of neutralization of SSPE (MUN-HT) and wild (EDM-W) strains of measles virus by rabbit antiserum against strain MUN-HT. (From Payne and Baublis, 1973, with permission).

has shown that the same defect of morphogenesis has been seen in SSPE brain in which tissue showed cell-fusion and in which many intracellular tubules were seen, but there was no sign of maturation of virions at the cell surface (Fig. 9.5) (Iwasaki and Koprowski, 1974).

HOST RANGE AND VARIABILITY OF ISOLATES

It is interesting that human embryo kidney and monkey kidney have been tried less frequently as sensitive media for growing SSPE virus than they were for isolation of wild virus from measles. For example, Horta-Barbosa used HeLa and HEp$_2$ cells in co-cultivation experiments with human brain cells and got evidence of virus antigen at the first and second passes in one case, but not until seven passes in the second case. Haemagglutinin was found along with giant cells and other cytopathic effects.

On further examination, one strain gave filterable virus, and produced much CPE and also haemagglutinin up to a titre of 1:8. The host range and plaque size resembled that of a vaccine strain (Edmonston, 749D) rather than that of a "wild" strain of measles virus, in that SSPE virus affected green monkey kidney cells, BS-C-1 cells, HeLa, dog kidney and chick embryo cells. Both strains were serologically less sensitive to the sera of the two patients than was wild virus, and both had equal sensitivities with "wild" measles against two measles-specific sera from children. The growth rate of the two strains was unequal, the poorly growing strain having less cytolytic effect (Horta-Barbosa et al., 1969b, 1970). In view of the finding that Horta-Barbosa's strain was cytopathic to chick embryo cells, it should be noted that Schumacher and his colleagues tested two SSPE measles virus strains and found no cytopathic effect on chick embryo cells, but only a limited infection and a very low yield of virus (Schumacher et al., 1972b).

Katz and his colleagues (1968) reported the induction of mild encephalitis in ferrets by intra-cerebral inoculation of non-cellular material from an SSPE brain, but they could not re-transfer encephalitis by cell-free extracts. Instead, explants of brain, not trypsinized, gave cultures containing giant cells and intracytoplasmic inclusion bodies. No free virus was found, but the cytopathic effect could be transferred to the WI$_{38}$ line of diploid human cells on which measles antigen could then be shown to be present by immunofluorescence and in which intranuclear tubules could be seen upon infection by each of two SSPE viruses (Katz et al., 1969).

A strain of measles, isolated from stored brain cells by co-cultivation with BS-C-1 cultured monkey cells (Payne et al., 1969), produced giant cells in the first culture, showed haemadsorption of monkey red blood cells only after twelve passes, and produced haemagglutinin about four passes later. Up till the time when specific haemadsorption developed, no free infectious virus was obtained. Six further passes gave a titre of no more than $10^{3\cdot3}$ infectious doses, but as much as eight units of haemagglutinin were present. In a later study it was noted that five strains of SSPE virus from different laboratories showed a great range of difference in properties of host-susceptibility, plaque size, and neuro-tropism but that all five were less susceptible than wild virus to neutralizing antibody (Payne and Baublis, 1973). This most careful study was accomplished by measurement of neutralization kinetics (Fig. 9.6 and Table 9.1) and it reveals the only constant difference between SSPE viruses and laboratory standard strains of measles virus.

TABLE 9.1

Lesser sensitivity to neutralization of SSPE viruses.
Neutralization of SSPE and other strains of measles virus by
convalescent-phase antiserum to measles, by the technique illustrated
in Fig. 9.6

Virus source and strain	Fraction of virus neutralized		
	Test	MUN-HT	EDM-W
SSPE			
MUN	0·38	....	0·73
BAX	0·34	....	0·78
GRI	0·42	0·33	....
GUS	0·43	0·43	....
SOW	0·44	....	0·72
Vaccine			
EMLP	0·88	....	0·82
SZ	0·89	0·34	....
Patient with measles			
EDM-W	0·85	0·44	....

(From Payne and Baublis, 1973, with permission).

The peculiar properties shown by measles virus when newly recovered from SSPE patients may be confined to virus isolated from brain. Two strains of virus were isolated by Horta-Barbosa from lymph node biop-sies taken from 5 patients (Horta-Barbosa et al., 1971). Fibroblastic cells from explants were used in conjunction with HeLa cells in spinner culture; one gave high yields of virus, $10^{8\cdot0}$ ID_{50} and 64 AD's, the

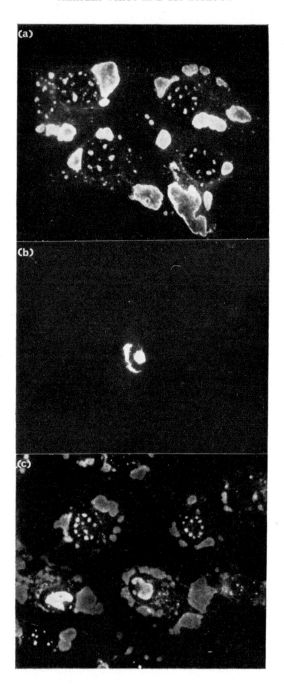

TABLE 9.8

Examples of variability in the properties of different SSPE measles virus strains.

Property	Observation	Reference
Growth—antigen synthesis	None, even with CPE—Rob strain.	Oyangai *et al.*, 1970.
	Rapid appearance of antigen in nucleus—Halle strain.	
	Slow appearance of antigen in nucleus—Dean strain.	Hamilton *et al.*, 1973.
Growth—virus	High—1st case $10^{8.0}$ $ID_{50/ml}$ Low—2nd case $10^{3.5}$ $ID_{50/ml}$	Horta-Barbosa *et al.*, 1971.
Haemagglutinin production	1 case—at 1st & 2nd pass 1 case—at 7th pass	Horta-Barbosa *et al.*, 1969b.
	1 case—at 16th pass	Payne *et al.*, 1969.
Cytopathic effect	Syncytia formed—Halle strain Pyknosis—Dean strain	Hamilton *et al.*, 1973.
Plaque size	Small	Horta-Barbosa *et al.*, 1970.
	Large	Payne and Baublis, 1971.
	Small changing to large	ter Meulen *et al.*, 1973.
Host range (in tissue culture)	Wide	Horta-Barbosa *et al.*, 1970.
	Restricted	Katz *et al.*, 1970b.
Neurovirulence	Greater than control (2 SSPE strains)	Lehrich *et al.*, 1970.
	Less than control (2 SSPE strains)	Albrecht and Schumacher, 1971.

FIG. 9.6. Patterns of fluorescent antibody staining in the nucleus of cells persistently infected with measles virus. (a) HEp_2 cells, persistently infected, treated by the indirect fluorescent antibody technique with guinea-pig anti-measles nucleocapsid. Most nuclei contain distinct and characteristic fluorescent aggregates. (b) SSPE. A section from SSPE brain treated by the direct method with fluorescein-conjugated serum from the patient. The two fluorescent masses are measles virus-specific. One is intranuclear. (Connolly, unpublished observations). (c) HEp_2 cells. The same culture as in (a), treated similarly to (a), but showing two intranuclear fluorescent masses resembling the appearance of inclusion bodies in SSPE.

other yielded virus after two passes at a titre of $10^{3 \cdot 5}$ ID_{50} and 2 AD's. The authors make the comment that the successful recoveries of virus were associated with early stages of the disease.

A list of observations on the varieties of behaviour shown by SSPE measles viruses at various stages of recovery and adaptation to laboratory conditions is best given in tabular form (Table 9.2). No factors are discernible in the accounts referred to which could influence the selection of the different properties manifested. In particular some marked differences occur between strains isolated in the same laboratories by the same methods.

SYNTHESIS OF VIRUS ANTIGEN

Additional evidence that SSPE virus has an abnormal growth cycle, at least in the early stages of recovery, comes from cytological studies using fluorescent antibody or peroxidase-conjugated antibody. Some of these have been carried out in conjunction with electron microscopic examination of the same tissues. The intranuclear inclusions seen in human brain are generally larger than those in measles virus-infected cells and the nucleus may be packed with virus antigen, unlike the circular spots and occasional aggregates which characterize carrier lines of measles virus and which are also seen in the late stages of lytic infection in tissue culture (Fig. 9.7) (Rustigian, 1962, 1966a; Rapp et al., 1960).

Measles virus-specific antigen is normally seen first in the cytoplasm of cells and only much later in the cell nucleus (p. 83). The fluorescent antibody studies by ter Meulen and his associates are of particular interest in showing the altered localization of antigen in cells infected with SSPE strains of virus. Inclusion body-like masses of antigen are seen, when SSPE sera or some early measles convalescent sera are applied in the indirect fluorescent antibody test to cells infected with these viruses, which are not revealed during infection of cells with non-SSPE strains of measles virus. The antigen was not identified, but the staining reaction was not blocked by antibody against virus haemagglutinin (ter Meulen et al., 1972a). A later study showed that SSPE virus produced intranuclear inclusions in all infected cells, as determined by the fluorescent antibody method, whereas wild viruses caused the appearance of intranuclear measles antigen in only one-fifth of infected cells and two days later than did the SSPE strains (ter Meulen et al., 1973; Tables 9.3, 9.4).

TABLE 9.3

Detection of intra-cellular antigen by indirect immunofluorescence in CV-1 cells

| | Measles | | | | | | SSPE | | | |
| | Edmonston | | Woodfolk | | Braxator[a] | | JAC | | LEC | |
	Cyto-pl.	Nu-clear	Cyto-pl.	Nu-clear	Cyto-pl.	Nu-clear	Cyto-pl.	Nu-clear	Cyto-pl.	Nu-clear
SSPE serum	+	−	+	+	+	+	+	+	+	+
Early measles convalescent serum	+	−	+	+	+	+	+	+	+	+
Late measles convalescent serum	+	−	+	−	+	−	+	−	+	−
Rabbit anti-measles serum	+	−	+	−	+	−	+	−	+	−

[a] Virus isolated from measles encephalitis.

TABLE 9.4

Comparison of different properties of SSPE and measles virus

	SSPE virus			Measles virus	Virus-measles encephalitis
		JAC	LEC	Woodfolk and Edmonston	Braxator
Virus production in tissue cultures	Low virus titre			High virus titre	High virus titre
Plaque formation	Heterogeneous small and large plaques			Uniform round plaques	Heterogeneous small and large plaques
Detection of viral antigen by immuno-fluorescence in infected tissue cultures.	Cytoplasm and nucleus. Intranuclear type of fluorescence: speckled or inclusion-like bodies.			Cytoplasm and nucleus.[a] Intranuclear type of speckled fluorescence.	Cytoplasm and nucleus. Intranuclear type of speckled fluorescence.
Electronmicroscopy	No granular filaments in virus particles			Granular filaments in virus particles	Granular filaments in virus particles
Animal experiment	Encephalitogenic			Not encephalitogenic	Encephalitogenic

[a]Edmonston—no nuclear staining. (Adapted from ter Meulen *et al.*, 1973, with permission).

Fluorescent antibody studies confirmed the asynchrony* of infection by SSPE virus that was also noted by Katz and his colleagues (1969). That is, antigen and nucleocapsid pile up in the nucleus and in nuclear inclusions before new virus is produced, whereas in lytic infection this nuclear accumulation is a late event. Secondly, these studies of antigen distribution point to a similarity between the growth of SSPE virus and measles virus replication in some carrier cell lines. Non-productive regulated infection gives similar asynchrony; in this case there is no question of neurotropism.

This virus–cell relationship of SSPE viruses is difficult to classify. ter Meulen states in a summary that cell-associated SSPE measles viruses caused encephalitis in all animals tested—hamsters, ferrets, dogs, calves and lambs (ter Meulen et al., 1973). Thus the unproductive infection in brain cells is neither abortive nor defective because it is transmissible to other cells; it is only partially suppressed. Various degrees of suppression can be shown to exist experimentally and to alter as passage proceeds. An SSPE measles virus grown in BS-C-1 cells from a biopsy specimen of brain taken from a 14-year-old boy revealed different relationships in different cell lines. One line, producing no infectious virus, was particularly interesting in that co-cultivation produced massive cell-destruction, but still yielded no infectious virus (Burnstein et al., 1974). It would be interesting to know whether non-yielding lines of this degree of incompleteness exist in vivo.

BIOCHEMICAL STUDIES ON SSPE VIRUSES

Only a few attempts have been made to compare the structural properties of laboratory strains of SSPE and measles virus.

The base composition of the 50S RNA of an SSPE strain has been reported by Yeh (1973) (see Table 9.5). The data suggest that there is a greater difference in the base composition of SSPE virus and measles than between measles, CDV, RSV, or SV5 as reported by other workers. It is unlikely that these minor differences are of great significance and their explanation probably lies in the relative amounts of plus and minus strands that are present in the RNA preparations studied.

Of more interest was Yeh's observation that the RNA from SSPE infected cells cross-annealed with measles RNA by only 60%. Recently, however, in a more extensive study Hall and ter Meulen (1976 and personal communication) have compared two measles strains (Edmonston and Woodfolk) with three SSPE strains (Mantooth, LEC and USC) by

* This is not the usual meaning of asynchrony in virus replication.

TABLE 9.5

Base composition of RNA from measles and related viruses

| Virus | Moles per cent | | | | Reference |
	A	U	G	C	
Measles	26·5	28·7	22·5	22·2	1
SSPE	24·2	31·1	21·4	23·3	2
CDV	27·8	30·1	21·1	20·9	1
RSV	26·7	31·0	21·9	20·2	1
SV5	27·6	30·6	20·5	21·5	1

(1) Bussell *et al.* (1974); (2) Yeh (1973).

competition cross-hybridization experiments in which care was taken to remove non-poly A containing RNA from both the radioactive and non-radioactive mRNA probes. Their results show that the Woodfolk and the three SSPE strains showed 100% homology with the Edmonston 50S RNA. However, in sharp contrast, when the comparison was made against the Mantooth (SSPE strain) only 90% homology was found with both measles strains, although the SSPE strains showed 100% homology. This information suggests that the SSPE strains may contain an extra piece of genetic information and Hall and ter Meulen (1976) have postulated that this may be derived by recombination with defective genomes of an animal virus, such as CVD or rinderpest, which could account for the rarity and predominant rural distribution of the disease.

At the time of writing only one report exists on the protein composition of SSPE virus. Schluederberg *et al.* (1974) reported that there was a small difference in the mobility of the membrane protein of the Edmonston strain and an SSPE strain. Although the difference detected was only 2,000 daltons, the measles virus protein was 38,000 and the SSPE was 40,000, the other major proteins had identical electrophoretic mobility. As the analyses were done by parallel runs on slab-acrylamide gels and were repeated up to nine times this small difference may indeed be significant. Although this report opens up another interesting approach to understanding the relationship between SSPE isolates and measles virus a more comprehensive study is required. However, Schluederberg's report is a further indication that the membrane protein may be involved in regulating persistent infection as discussed on p. 112.

NEUROPATHOGENICITY

Chronic encephalitis

Encephalitis was successfully transmitted from SSPE brain to ferrets before the causative agent was recognized as measles (Katz *et al.*, 1968, 1970a) and since then a series of important reports from different laboratories describe the results of inoculating SSPE measles virus into animals. Dogs and ferrets both got symptoms of chronic infection of the brain following intra-cerebral inoculation of brain biopsy homogenate containing the LEC strain of SSPE virus. The former also got signs similar to "hard pad disease" of distemper, a related virus disease (Tijl *et al.*, 1971; Notermans *et al.*, 1973). Calves and lambs also showed clinical signs of encephalitis following the injection of LEC virus in the cell-associated form within CV-1 cells (Thein *et al.*, 1972). Cell-associated virus is by far the most effective inducer of chronic encephalitis. It is of great interest that all these animals, which would have developed antibody and sometimes symptoms if they were inoculated with live measles virus, did have symptoms but developed no measles-specific antibody as a consequence of infection with the SSPE virus. This was the experience summarized by ter Meulen *et al.* (1973), but it is not a constant finding, being contrary to the experiments of Katz and his colleagues with ferrets which developed low titres of anti-haemagglutinin after the intra-cerebral inoculation of brain tissue from SSPE (Katz *et al.*, 1970a). Clearly humoral measles-specific antibody is not an essential mediator for the production of chronic encephalitis by SSPE virus. In some of the affected animals lesions are very similar to those of SSPE, including the production of typical Cowdry-type intranuclear inclusion bodies (Katz *et al.*, 1970a) which are also sometimes seen in cultured SSPE virus (Oyanagi *et al.*, 1971).

One difficulty which arises if little or no infectious virus is produced in SSPE is to account for the wide distribution of virus-infected cells in the brain. The problem has been partly solved by the demonstration of cell-fusion in the cerebrum between cells that could be shown to be packed with typical nucleocapsid tubules (Fig. 9.5) (Iwasaki and Koprowski, 1974). Similar breaks have been seen in virus-infected cultures of organized neural tissue (Fig. 9.4) (Raine *et al.*, 1973) and cell-to-cell spread of virus is thus accounted for. Whether virus structures are required at the surface of the cell to promote this fusion is not easy to ascertain, but it may be noted that although Iwasaki and Koprowski (1974) saw no sign of budding or differentiation of cell-membrane in sections of cerebral tissue, the same material examined in suspension by the negative-staining method, did show pieces of fringed membrane with adherent

virus tubules. Small and widely scattered patches at which some or all
of the maturation process was taking place would not easily be seen by
electron microscopy but would easily promote junction between cells.

The ability of SSPE viruses to initiate chronic encephalitis in species
of adult animal is not matched by any similar property of non-SSPE
measles viruses. Whether it is a true stable property which does not
disappear as the strains become adapted to transfer by extracellular
infectious particles does not seem to have been tested adequately. Since
the chronic encephalitis can most easily result from the inoculation of
cell-associated, less productive, SSPE virus (Tijl et al., 1971; Thein
et al., 1972; Notermans et al., 1973), it may be that the less powerful
virus-immune response given by such non-productive inoculum allows
infection to become established and likewise fails to eradicate it. Partial
passive immunity in weanling hamsters has been used to allow the
establishment of prolonged infection of hamster brain by a neurotropic
strain of measles virus (Wear and Rapp, 1971) and suppression of the
immune response can convert latent into overt encephalitis (Albrecht
et al., 1972; Wear and Rapp, 1971) which makes the role of the immune
state of the patient or experimental animal a critical one in SSPE or
its laboratory counterpart.

Three significant papers on the subject of the pathogenesis of experimental measles encephalitis caused in hamsters by a neuro-adapted
SSPE virus, HBS, have recently been published.

Chronic infection was observed in hamsters which had survived the
acute disease (p. 121), there was demyelination in most lesions and no
budding of virus was seen in virus-infected cells (Raine et al., 1974b).
The effect of immunosuppression by anti-lymphocyte serum was sufficient to convert a non-lethal encephalitis to one carrying a 50% mortality in which the immunosuppressed animals carried much cell-free
virus. Partial or transient immunosuppression could incite a subacute
but fatal encephalitis accompanied by the recovery of cell-associated
virus or by no recovery, but brain cells contained much virus antigen.
The parenchymatous lesions consisted of foci of necrosis, but ependymal
lesions consisted of small multinucleated cells (Byington and Johnson,
1975). The third paper illustrates the difference between the acute disease
in newborn, as opposed to weanling, animals. Neurones of the newborn
are affected and the morphogenesis of virus is complete, with capsid
alignment and budding of virions; in the weanling occasional viruses
were seen up to the eighth day of infection. Thereafter they were rare;
there was no sub-membranal alignment of nucleocapsid and no typical
viruses were produced (Raine et al., 1975). The influence of immune-
competence on the outcome of infection and the association between

age and susceptibility, presumably in relation to immune function, is well seen in these experiments. It is considered also on p. 123.

Acute encephalitis

Encephalitis is not a common sequel to measles. The prolonged association between SSPE virus and brain cells raised the question of whether SSPE viruses arose from neuropathogenic strains of measles virus. Neuropathogenicity as tested by intra-cerebral inoculation into hamsters or mice, was therefore the second important characteristic of SSPE strains of measles virus to receive attention. Their neurovirulence varies from strain to strain and conclusions cannot be clearcut because of the existence of laboratory strains of measles viruses which were fully virulent for young rodents without adaptation to brain, when tested in the same system as a fully virulent SSPE strain of virus (e.g. Katow et al., 1973). With this in mind, we note the two SSPE strains were more neurovirulent than two wild-type measles viruses, when tested in suckling hamsters (Lehrich et al., 1970); another was much more encephalitogenic than a wild strain of virus in newborn hamsters (Cernescu et al., 1972). Katow and his colleagues had a non-yielding SSPE strain Niigita which was almost wholly lethal for infant hamsters and mice whilst other strains were innocuous and yet others, virus releasing strains grown in Vero cells in the same laboratory, were quite virulent (Shishido et al., 1973). In contrast, Albrecht and Schumacher (1971) record that the Mantooth strain of virus and Ro-SSPE-1 virus, after a limited number of passes in Vero and BS-C-1 cells respectively, were lethal to only 5% or 6% of inoculated mice, as compared with the 21% to 39% lethality of inoculated Edmonston virus and the 68% to 86% lethality of the Schwarz chick-cell attenuated Edmonston virus that had been passed only twice in Vero cells. These variations and the fairly rapid alterations of virulence, produced by changing cultural circumstances, of other strains of measles virus, make the different degrees of neurotropism of SSPE strains fairly meaningless. Much more work is required on standardization of methods and on study of the many factors which determine neurotropism of virus given by various routes in different species of animal.

There are few studies on the changes in neuropathogenicity of SSPE strains of measles virus which take place as they become adapted to laboratory culture but a Roumanian strain, which was reported by Cernescu et al. (1972) to be more pathogenic for infant hamsters than a wild strain of measles virus, Co 69, was found to lose neurovirulence after five passes in WI_{38} human embryo lung cells and to yield less virus intracerebrally (Draganescu et al., 1972). These experiments were complicated by infection of the test species with ectromelia virus. As for deliberate adaptation to brain, Castro *et al.* (1972) are of the opinion that Mantooth SSPE virus, adapted to hamster brain, retained all the characteristics of standard measles virus, namely, infectivity, plaque formation, typical cytopathic effect, inclusion body formation, haemagglutinin production, haemadsorption and fluorescent antibody pattern of intracellular distribution of antigen.

Thus we see that neuropathogenicity is not particularly characteristic of SSPE viruses, or rather, the study of neurovirulence in SSPE measles strains is subject to the same limitations which apply to the study of various other characteristics, none of which has been considered as a suitable genetic marker for isolates of measles virus.

COMMENT

SSPE viruses are serologically measles virus; they produce similar cytopathic effects, agglutinate monkey erythrocytes and grow in the same hosts as does measles virus. Early in the course of recovery and adaptation to tissue culture they differ particularly in behaviour from laboratory strains of measles virus. The differences are as follows:

1. There is a stronger association than normal between SSPE strains and the cell nucleus which has been sometimes described as asynchrony of replication because it reverses the sequence of events usually seen in lytic infection by measles virus.

2. Infection tends to be non-productive in the early stages of propagation of SSPE virus-infected cells and the presence of virus is indicated by a cytopathic effect, by the presence of measles virus antigen in cells of the culture, or by detecting virus substructures microscopically.

3. Success in recovering SSPE virus is variable but, when successful, full reproductive ability is restored progressively with passage.

4. Alternatively, a permanent non-productive regulated infection may be established in which cytopathic effects unaccompanied by the production of free virus may occur spontaneously or after co-cultivation of the culture with measles virus-susceptible cells. This important observation implies that the cell-virus complex is pathogenic towards adjacent cells.

5. The formation of syncytia, seen in virus non-yielding cultures of SSPE viruses has its analogy in cell-to-cell transmission through fused membranes of neighbouring cells in the brain.

6. Morphogenesis of virus is halted or delayed in brain and in tissue culture at the stage of virus maturation. Nucleocapsids, though abundant, are not brought forward to become incorporated into cell membrane

and form viruses. The reason for this defect is not known, but is not due to lack of differentiation of the cell-membrane.

7. SSPE virus, unlike laboratory-adapted strains of wild measles virus, can, when inoculated in the cell-associated and less productive state, induce chronic encephalitis in ferrets, dogs, calves, lambs and hamsters, without subsequent production of measles virus-specific antibody.

8. The strains of SSPE viruses which have been compared with measles virus in neutralization-tests against natural measles antibodies have been a little less sensitive to the antisera. This applies also to serum taken from SSPE patients. Similarly, less sensitive strains of measles virus have been recovered amongst survivors of plaque-neutralization tests, using wild virus.

9. SSPE viruses vary in their ease of isolation, growth rate, host-range, plaque size, pattern of fluorescent antibody staining, type of haemagglutinin and yield of infectious virus, much like other measles viruses.

10. Two reports refer to minor differences in the RNA and protein of SSPE viruses as compared to those of measles virus, but they are unconfirmed yet.

No single stable character of SSPE viruses has been found which is not present in some preparation or other of standard measles virus, with the possible exception of lesser sensitivity to neutralization by immune sera. Taken with the marked tendency of SSPE strains to acquire the character of standard measles virus on continued propagation it is likely that the opinion of Hamilton and Dubois about the adaptation of their own two SSPE virus strains should apply to all: "once released from constraint by co-cultivation all the properties of measles virus return".

The nature of the molecular events which explain how the whole genome of the measles virus can survive in static neurones in SSPE, can be transmitted when brain cells are allowed to multiply and gradually assume all the replicative functions leading to the production of mature virus remains unexplained. Similar relationships between virus and cell are seen in artificially induced, persistently infected cell cultures and the same molecular mechanisms are likely to operate in these and *in vivo*. The possible influence of defective particles is mentioned on p. 111 but their existence in SSPE is hypothetical. Likewise, integration of a DNA copy of measles virus RNA into the infected cells of SSPE brain, as into persistently infected cell cultures (Zhdanov, 1975; Simpson and Iinuma, 1975) has not been demonstrated in brain.

10

Immunology of measles virus—epidemiology; specific antibodies

INTERPRETATION: (1) *The epidemiology of measles is closely related to immunity, as measured by circulating antibody.* (2) *Antibody to capsid, haemagglutinin and haemolysin are known. Virus neutralization is best associated with anti-haemolysin.* (3) *Measles-specific antibody lasts a lifetime and declines very slowly after a preliminary moderate fall.* (4) *Measles virus-specific IgM does indicate active or recent infection by measles virus and has not otherwise been found except in multiple sclerosis.* (5) *Some measles antibodies are cytotoxic for measles virus-infected cells and antibodies may exist which also block the cytotoxic effect.* (6) *The functions of each antibody remain to be defined as does the effect of each on the control of the total immune response to measles virus.*

It may be thought that the immunology of measles virus and of measles is a subject concerned only with human immune responses in which the part played by this very stable virus, as antigen, is wholly passive and neutral, much like that of any inert protein that may stimulate immuno-competent cells. This is not so. Some old and some new observations suggest strongly that there is a fundamental relationship between living measles virus and some population of immunocompetent cells which remains to be discovered and which, when found, ought to explain, amongst other things, the long-lasting solid immunity which follows

measles, the diminution or abolition of the del
reaction during measles, the palliative effect of
conditions, para-infectious measles encephalitis,
antibody responses in certain pathogenic states wh
measles, abnormal sensitivity of immunosuppresse
vaccines and abnormal severity of measles in m:
sections have been written with these points in m

162

 ## EPIDEMIOLOGY AND IMMUI

Some important facts about immunity against measles are common
knowledge. Prevalence of the disease in childhood, the rarity of second
attacks, its constant presence in large populations and its periodic seasonal incidence, are all taken for granted. Almost as well known is the
fact that measles may disappear from isolated communities for long
intervals and, when it is reintroduced into them, epidemics of great
intensity and abnormal severity are apt to occur. It is less well known
that, in such attacks as do occur after a long episode of freedom from
measles, the disease affects only those, and nearly all of those, who have
not experienced it before. For example, in Greenland in 1951–1954, only
five out of 4,262 susceptible individuals escaped infection, nearly half
of the patients developed complications and the mortality rate was 18
times higher than that in a subsequent epidemic (Bech, 1962).

The immunity of those who escape an attack when again exposed to
measles, is long-lasting, more than 50 years (Panum, 1847), even when
virus has been absent from the community, a point which is of some
significance in attempting to understand the nature of this long-lasting
acquired resistance. Many surveys of measles-specific antibody in populations support the idea that the waning of specific immunity with time is
unusually slight, in the circumstances. It also follows from those well-
established observations on isolated communities, that those who have
recovered from the infection do not disseminate it afterwards; measles
is neither apparent nor, in its absence, is natural immunity ever acquired.
It also follows that, if measles were ever eradicated universally by vaccination, it could not return.

Freshly caught monkeys which are susceptible to measles, form an
excellent, though inadvertent, indicator of the constant presence of
measles in a community, by becoming rapidly converted to the immune
state. This need not mean direct contact with an overt case of measles,
for Stokes has shown that, provided antibody titres are less than 1 in 8
when human re-exposure to measles takes place, a significant rise in
titre sometimes occurs, indicating that sub-clinical infection and, presumably, temporary carriage are possible (Stokes et al., 1961).

portal of entry into the body is almost certainly respiratory but experiments (Papp, 1956) and clinical data on keratitis (Thygeson, 1959; Enders, 1962; Florman and Agatston, 1962) have cast suspicion on the conjunctiva as a likely second receptive area for the virus. The seasonal incidence may be related to climatic conditions or to social habits secondary to these, it is not known which, but at least measles morbidity and the physical resistance of the virus are similarly correlated to a low relative humidity (de Jong and Winkler, 1964; de Jong, 1965). Otherwise the behaviour of the virus in nature is conditioned almost entirely by herd immunity. A proportion of 20% of susceptible individuals in a population maintains the virus at endemic incidence, a porportion of 25% is said to be sufficient to initiate a seasonal epidemic (Babbot and Gordon, 1954, quoting Brincker, 1936).

In scattered populations the mean age of onset is later than in urban areas and this has been related hypothetically to the rural incidence of subacute sclerosing panencephalitis, a complication of measles (Detels et al., 1973).

TESTS FOR ANTIBODY

Neutralizing and complement-fixing antibodies to measles virus were the first to be described (Enders and Peebles, 1954). With the discovery of virus haemagglutination and haemolysis came anti-haemagglutinin and anti-haemolysin tests (Periés and Chany, 1960) and the separation of the membrane-bound haemagglutinin–haemolysin complex from the internal nucleocapsid revealed that at least half of the complement-fixing antibody in a serum was in fact reacting with the internal component (Waterson et al., 1963). The same antigen is also the main one reacting with antibody in gel-diffusion tests (Norrby and Gollmar, 1972).

The physiological function of each of these antigen–antibody reactions is not completely known, but neutralizing antibody combines with the envelope complex, especially antihaemolysin, whilst capsid antibody is non-neutralizing (Norrby and Gollmar, 1975). Antibody to measles can be toxic to measles virus-infected cells (Minagawa and Yamada, 1971; Kibler et al., 1974; Joseph et al., 1975a).

Recently Norrby and his colleagues have found evidence that the anti-haemolytic effects of human sera may be separated into a non-anti-haemagglutinating part and an anti-haemagglutinin which, by counteracting adsorption of the virus to the red cell surface, has also got an anti-haemolytic effect. On testing human measles-convalescent sera with the non-haemolytic Tween-80-ether-treated haemagglutinin and also with whole virus haemolysin, they first detected one serum which had a higher anti-haemolytic than anti-haemagglutinating titre (Norrby and Gollmar, 1972). More tests on sera from which anti-haemagglutinin had been absorbed with non-haemolytic, ether-treated virus, revealed residual anti-haemolysin which was positively correlated in amount with the neutralizing titre of the serum (Norrby and Gollmar, 1975). Further, sera prepared in rabbits by injecting spikeless, trypsin-treated virus particles, had low titres of anti-haemagglutinin, but higher titres of anti-haemolysin (Norrby and Gollmar, 1975). A consequence of this was the clinically important observation that antibody which developed after a vaccination with live measles vaccine contained the kind of anti-haemolysin which is independent of anti-haemagglutinin. Antibody formed after immunization with killed vaccine contained only the anti-haemolytic property which is linked to anti-haemagglutinin (Norrby et al., 1975).

In diagnostic work anti-haemagglutinin was found at first to provide a more sensitive test than neutralization and complement fixation was least sensitive. Virus, disaggregated by treatment with the detergent Tween-80 and ether, but not by ether alone, gave higher haemagglutinin titres (Norrby, 1963; Waterson et al., 1963) and, therefore, dose for dose revealed higher antibody titres than did whole virus (Enders-Ruckle, 1965). More sensitive cultural systems and the plaque assay have not reversed the sensitivity of the tests but passive haemagglutination is more sensitive than the anti-haemagglutinin and complement-fixation methods (Barron et al., 1963; Cutchins, 1962; Kenney and Schell, 1975). Sensitivity refers here only to the relative titres found in the serum, not to the amount of virus detected. Fluorescent antibody staining of virus-infected cells in culture provides a titration which, in our own experience, is about equal in sensitivity to the anti-haemagglutinin test. Antibody to virus components at the cell surface, in the cytoplasm and within the nucleus can be separately titrated in this way, but not yet correctly identified, until complete definition of the antigenic specificity of the staining pattern becomes available, though measles virus antigen on or in the nucleus seems to be mostly nucleocapsid (Norrby, 1972) and anti-haemolysin and anti-membrane protein give different staining patterns (Gharpure and Fraser, unpublished data).

APPEARANCE AND DURATION OF CIRCULATING ANTIBODY

Measles antibodies, tested by any means, may first be detected from the first to the third day of the rash; they reach their maximum titre by the twenty-fifth day (but usually before then) after the onset of illness (Enders-Ruckle, 1965). After a slight drop in titre during the next 8 to

18 months or so (Bech, 1960a, c; Norrby *et al.*, 1966), they persist for life, with a slow decline for more than 30 years after the illness (Ruckle and Rogers, 1957; Black, 1959a; Dekking, 1965). Using the rather insensitive complement-fixation test, Dekking found tests to be positive even at 80 years of age in 15% of individuals, and noted that this test, unlike neutralizing antibody (Stokes *et al.*, 1961), did not rise in titre on mere re-exposure to measles. A most valuable general picture of declining antibody titre with time and in relation to environmental exposure is given by Krugman (1971; Figs. 10.1 and 10.2).

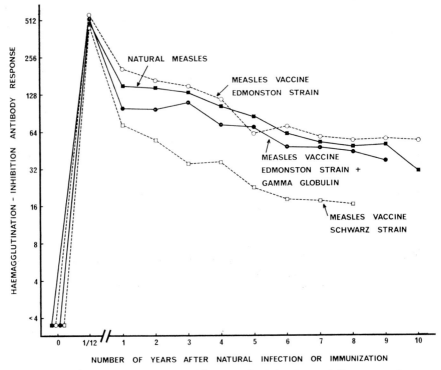

FIG. 10.1   Geometric mean anti-haemagglutinin-titres following infection or vaccination with live attenuated measles vaccines. (Figure from Krugman, 1971, with permission).

Susceptibility to measles seems to exist in those who have neutralizing antibody titres of less than 1:8 (Stokes *et al.*, 1961; Karelitz and Markham, 1962).

In commenting on reasons for this long-lasting measles-specific antibody, Enders-Ruckle (1965) considers that there are only two possible mechanisms which could provide an explanation. One is re-infection by the same or by cross-reacting viruses, the other is persistence of infection accompanied by the production of virus or of sub-units of the virus. In support of the second possibility she quotes the isolation of four virus strains recovered from the lymph nodes of two patients who had had measles several weeks before. Measles virus of course is frequently recovered from monkeys (Ruckle, 1958) and has been found in one culture of human kidney cells (Melnick et al., 1965) so that persistence in vivo is a fact. The problem in adult man has not been settled and we describe

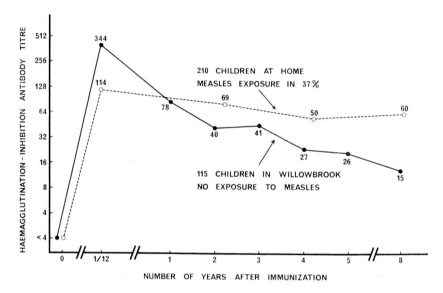

FIG. 10.2. Comparative geometric mean antibody titres of children living at home or in an institution after being vaccinated with the same attenuated vaccine. (Figure from Krugman 1971 with permission.)

the serological observations because of the importance of the question of latent infection to the aetiology of chronic measles panencephalitis in which antibody titres are higher than average (Connolly et al., 1967; Freeman et al., 1967) and also because antibodies against measles virus are slightly increased in multiple sclerosis (Adams and Imagawa, 1962) in diseases of the liver (Closs et al., 1971, 1973; Triger et al., 1972) and in systemic lupus erythematosis (Hollinger et al., 1971; Kalliomaki and Halonen, 1972; Laitenen and Vaheri, 1976; pp. 190, 194).

CLASSES OF SPECIFIC IMMUNOGLOBULIN

During active infection measles virus, like other infectious agents, stimulates the production of specific IgM which is normally absent from late convalescent sera and from adult serum (Haire, unpublished data; Connolly *et al.*, 1971). Haire's findings follow and conform to physical data on the presence of virus-specific 19S and 7S globulins (Schluederberg, 1965) and are probably more sensitive, since a circulating life of only one month was detected by physical methods for measles virus-specific IgM (Tikhonova *et al.*, 1973). The duration of measles-specific IgM estimated as 19S globulin by Karaseva *et al.* (1974) is 8 to 10 weeks. Some patients with subacute sclerosing panencephalitis have a measles-specific IgM in serum and in CSF (Connolly *et al.*, 1971; Khozinsky *et al.*, 1974; Thomson *et al.*, 1975) which is understandable in view of the amount of measles virus antigen in the affected brain and the amount of specific antibody found in cerebrospinal fluid. Some patients who have multiple sclerosis have a measles-specific IgM in serum though not in

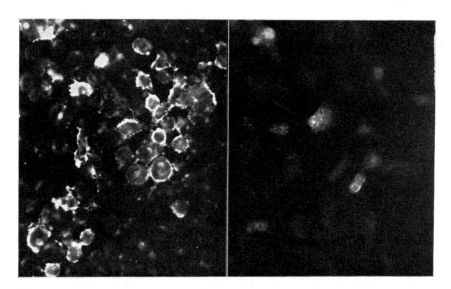

Fig. 10.3. The acetone sensitivity of fluorescent antibody staining by MS–IgM. Pretreatment of measles virus-infected HEp$_2$ cells with acetone prevents union of infected cell membrane with MS-IgM. (Left) Unfixed, infected culture treated with MS serum followed by FITC-conjugated anti-human IgM. Fluorescence indicates that IgM is bound. (Right) A sample from the same culture, fixed 10 min in acetone at room temperature, then treated similarly. MS–IgM is not bound.

cerebrospinal fluid (Haire et al., 1973b) and no measles virus has ever been shown to be present in patients with this disease. The receptor for the IgM in MS sera is distinctly acetone-sensitive (Fraser et al., 1974; Fig. 10.3). Not much is known about how the IgM antibody response to it is controlled nor what it may signify but it is an anti-haemolysin.

Measles virus-specific IgG has been found to increase significantly in two patients who were being treated with immunosuppressive drugs (Fiala et al., 1975). It is not known whether this represented re-activation of a latent infection or whether it resulted from re-infection. Linneman (1973) found that a second infection occurring in patients with measles-specific antibody did not induce the production of measles-specific IgM in detectable amounts, but whether this holds true during immuno-suppression is not known.

Measles virus-specific IgA also appears in serum in response to infection (Tikhonova et al., 1973). Bellanti noted that live measles vaccine stimulated an IgA response in 7 out of 9 children, as opposed to only 3 out of 7 given killed vaccine, as one would expect from a vaccine that was able to reach and replicate in organs containing IgA-producing cells (Bellanti et al., 1969). One attempt to look at the influence of measles-specific IgA on the outcome of measles in African children could find no correlation between the two (Geddes and Gregory, 1974).

ANTIBODY AND INFECTED CELLS

Measles-specific antobodies are cytotoxic to measles virus-infected HeLa cells in the presence of complement and distinct from anti-cellular cytotoxic anti-globulin in the same serum (Minagawa and Yamada, 1971), for the latter antibody seems to be more complement-dependent and has a much more rapid cytotoxic action than do measles virus-specific antibodies. Virus-specific cytotoxic antibody is characteristic of the antibody response in early measles, in atypical attacks of measles and in subacute sclerosing panencephalitis, all virus-active conditions. It seems to be different from anti-haemagglutinin in that it is not removed from serum by absorption with virus haemagglutinin, nor is the neutralizing antibody, but cytotoxic antibody does seem to be dependent for its action on the presence and nature of virus antigen at the surface of the infected cell. Thus, a carrier line of the LEC strain of SSPE measles virus which carried little virus at the cell surface, is stated not to react with this anti-body (Kibler et al., 1974; Kibler and ter Meulen, 1975). Cytotoxic anti-bodies exist which are specific for other viruses, e.g. cytomegaloviruses, and have been used as control systems in testing measles infected cells (Thurman et al,. 1973).

Studies of measles-specific cytotoxicity by Ehrnst (1975) showed that the toxic globulin was IgG rather than IgM and that it was present also in the cerebrospinal fluid of patients with subacute sclerosing panencephalitis and of many with multiple sclerosis. However, Ehrnst, unlike Kibler and his colleagues, tended to find that cytotoxicity and anti-haemagglutinin were equated in measles virus-specific rabbit antibody. Although the antigenic component of measles virus which reacts in the cytotoxic tests remained unidentified, the sensitivity of the infected cells was observed to decrease in culture from day to day and when cytotoxicity was at its height, fluorescence staining showed that antibody was distributed at one or other pole of the cell (Ehrnst, 1975). Prozoning was noted, suggesting the existence of blocking factors in serum.

Joseph and Oldstone (1974) have described what seems to be redistribution of measles virus antigen on the surface of carrier cells which were exposed to measles virus-specific antibody. They too found that measles-specific IgG was cytotoxic and, further, that the specific antigen–antibody complex at the cell surface activated the complement system by the alternative route from C_3 (Joseph et al., 1975a). It seems possible that this finding is connected with Minagawa's observation that virus-specific and cell specific antibodies have different dynamics (p. 167). As free measles virus combines with specific antibody and, in so doing, activates the classical complement pathway, the involvement of membrane-bound antigen with the alternative pathway may be of importance in pathogenic states in which infected cells continuously bear measles virus antigens at their surface. The probable existence of blocking-antibody to cytotoxic serum factors (above) may well account for some divergence of results in different patients as it may do in tests of cellular cytotoxicity (Ahmed et al., 1974).

COMMENT

The uniformity of the community response to measles and the lack of naturally occurring serological variants of measles virus may have suggested in the past that a very simple equation might explain the relationship between susceptibility, disease and acquired resistance in man's encounter with this organism. Analysis of the antigens of the virus and of the antibodies in sera have revealed that the apparent simplicity must be the end result of a complex of interactions. Common sense would suggest that although anti-haemagglutinin and anti-haemolysin may act against different antigens on the surface of the virus, their harmonious combined actions are unlikely to be antagonistic to neutralization. The predominant antigenic influence of haemolysin in stimulating protection, as described by Norrby, is of great practical consequence. What is not yet clear is how the effectiveness of neutralization may be upset by disproportion between the amounts of each kind of antibody present. For example cytotoxic antibodies cannot play a wholly beneficial role, especially if the virus-infected cells happen to be in a vital organ. The reason why activation of complement should occur by two different pathways, according to whether virus antigen is free or bound to cell membrane, is incomprehensible at present. It would be unwise to assume that redistribution of measles virus antigen on the cell membrane and subsequent stripping of antigen from the surface, as seen *in vitro* (Lampert *et al.*, 1975), has no counterpart *in vivo* and the nature of blocking anticytotoxic factors in serum needs to be explored. It is not known whether they encourage or oppose persistent infection by measles virus. Finally, the nature of interplay between measles virus-specific antibodies and cell-mediated responses is almost unknown.

II | Immunology of measles virus—cell-mediated immunity

> INTERPRETATION: (1) *Measles causes a leukopoenia and a depression of certain cell-mediated reactions.* (2) *It is not known whether these two events are functionally connected.* (3) *Measles virus in cultured lymphocytes depresses immunological responses and also responses to cell mitogens. Some of the effect is due to non-viral products of infection.* (4) *Specific responses to measles antigen are less easy to measure, but responses to crude measles preparations certainly occur and specific lymphocyte cytotoxicity towards virus-infected cells has been shown.* (5) *Information about the effect of measles on the functioning of sub-populations of lymphocytes and on other mononuclear leukocytes is badly needed.*

Although studies of cell-mediated immunity are of great consequence to recovery from, and resistance to, infection, systematic studies have not been made to any extent that is comparable with antibody surveys. Yet there have long been two clinical indications of the significance of cell-mediated immunity to measles. The older is the observation that delayed hypersensitivity, as recorded by the Mantoux-type of intra-dermal test, is greatly depressed or even absent during measles (Pirquet, 1908); the more recent is the knowledge that certain kinds of immuno-deficiency which leave the patient unable to make antibody do not prevent normal recovery from measles (Good and Zak, 1956). Invasion of the lymphoid and reticulo-endothelial cells is such a profoundly important part of the pathogenesis of measles that the first observation is not unexpected. The other observation was one of the first indications to be obtained that cell-mediated immunity is mainly responsible for recovery from most viral and many bacterial infections.

CIRCULATING LEUKOCYTES AND MEASLES VIRUS

A leukopoenia during measles in man or in monkey is not the most constant clinical finding, but it is an established characteristic which occurs most markedly about the 11th to 13th day after exposure to the infection*. The administration of live measles vaccine gave to Black and Sheridan (1967) an opportunity to follow the cellular changes in some detail and they reported that a "broad trough of leukopoenia was apparent" which began about the 4th day after infection, gave a minimum mean count about the 11th day and ending, in their subjects, who were 39 children, at the 13th day after inoculation. This type of leukopoenia occurs also in monkeys (Blake and Trask, 1921b; Taniguchi et al., 1954b; Sergiev et al., 1960), but is is not pathognomonic of measles, for remarkably similar effects were seen to follow artificially induced hepatitis (Havens and Mark, 1946). The declining white cell count sometimes affects lymphocytes on the 3rd day and neutrophils mostly about the 8th day after infection, whilst monocytes are hardly affected, but eosinophils to all intents and purposes disappear from the circulating blood (Fig. 11.1a,b). It is unfortunate that these changes have not been taken into account in many of the tests of lymphocyte function that have been made during an attack of measles, a fact referred to in correspondence by Foreman (Kadowaki et al., 1970).

During the viraemic phase of measles, leukocytes are a source of virus (Papp, 1937; Gresser and Chany, 1963; Peebles, 1967) and Osunkoya has lately shown that the carriage of virus is passive, not replicating until the leukocyte is stimulated by phytohaemagglutinin or by mixed cultures with other lymphocytes. In these circumstances leukocytes develop giant cells and synthesize measles antigen, as seen by the fluorescent antibody technique, which they do not do if unstimulated (Osunkoya et al., 1974a and b). The association between cell stimulation, cell division and virus replication has been observed in cultures of continuous cell lines (Karaki, 1965b) and certainly greater numbers of human leukocytes infected in vitro synthesize measles virus antigen when stimulated mitogenically than when unstimulated (Barry et al., 1976; Shirodaria, unpublished data).

* For older references and detailed differential counts see Benjamin and Ward (1932).

Detailed examination of sub-populations of human leukocytes has revealed that measles virus readily grew and persisted in lymphoblast lines of the T type and the B type. Freshly prepared leukocytes from circulating blood also supported the growth of measles virus more readily after stimulation with mitogen, about 6% to 8% of T and B lymphocytes and 10% of macrophages being susceptible. The yield of infectious virus per lymphocyte was 10–20 pfu, at least ten-fold less from infected monocytes, and virus did not grow in polymorphonuclear granulocytes (Joseph et al., 1975b).

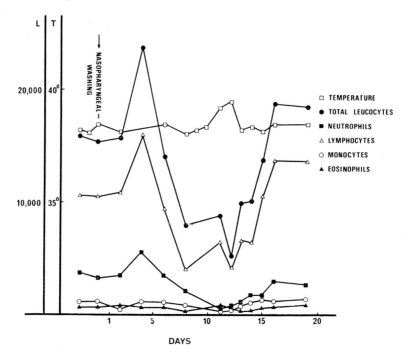

FIG. 11.1   Total and differential leukocyte counts on a monkey inoculated with 1 cc nasopharyngeal washing from patient, taken one day after eruption. (Slightly adapted, with permission, from Taniguchi et al., 1954b.)

NON-SPECIFIC RESPONSES FOLLOWING MEASLES IN VIVO

Cell-mediated responses are, and well might be, altered during infection by measles virus which not only depresses the numbers of several kinds of circulating leukocytes, but also grows in a proportion of them. Pirquet's (1908) observation that measles inhibits the dermal tuberculin reaction has been confirmed many times*. (Mitchell *et al.*, 1935; Pilcher, 1935; Bentzon, 1953). The availability of live vaccine and the advent of *in vitro* tests of lymphocyte function made it possible to re-examine under defined conditions the suppressive effect of measles on the Mantoux and other intradermal tests of delayed hypersensitivity, as well as to test the response of leukocytes to antigens or mitogens, once placed *in vitro*. The results are not clear cut, but they are at present a little more consistent than tests of various specific cellular responses to measles virus, in health or disease. The main effect is clearly one of suppression or inhibition of normal immunological activity (Table 11.1).

A controlled test of the effect of live measles vaccine on hypersensitivity to purified protein derivative of tuberculin was provided by children given measles vaccine alone, or vaccine along with a small dose of measles-immune human globulin. Of 7 children given vaccine only, 6 had a reduced response to the tuberculin preparation; 10 children given vaccine along with antibody showed no reduction of the same response. This confirms the old data on the effect of measles and reveals that depression of the response can be counteracted by circulating antibody (Mellman and Wetton, 1963).

There are practical consequences of the *in vivo* immuno-suppressive effect of measles. It is probable that increased susceptibility to tuberculosis, long postulated, is founded on fact (Bech, 1962). In addition, humoral immunity may be adversely influenced; Askerov and Voroshilova (1968) compared the effects of measles and of measles live vaccine on the development of antibody to poliovirus. For 3 months after infection the titre of specific poliovirus antibody in the blood remained below normal levels. The decline was less when passive immunity to measles was present, 'and it did not occur after the infection by attenuated virus.

Live virus is required to produce the foregoing effects. Using a cytological assay (blast transformation) and skin testing, Fireman carried out a small, but thorough comparison of the influence of live, with that of killed, measles vaccine on delayed hypersensitivity, lymphocyte responses and immediate hypersensitivity to various antigens. The skin reaction to PPD, candida albicans, diphtheria toxoid, and vaccinia virus was reduced, but not lost, in all subjects given live vaccine; dead vaccine, which did not affect the response to PPD, did slightly reduce the response to diphtheria toxoid. Two weeks after vaccination with live virus, the *in vitro* cellular hypersensitivity to antigens was clearly reduced, but the

---

* There are contrary results according to technique of testing, e.g. Nalbant (1937), but Pirquet's careful observations clearly indicate maximum inhibition at the first 6 to 8 days after the onset of the rash in his series of 24 individuals.

TABLE 11.1

Effect of measles in vivo on lymphocyte stimulation

Reference	Measles Virus Present as:—	Delayed Skin Test	Results of Lymphocytic stimulation by			Assay	Medium
			PHA	PPD	Other		
Mellman and Wetton (1963)	Live vaccine and with gamma globulin control	PPDS at 48h reduced and twice weekly. 6/7 reduced; 0/10 reduced when gamma globulin given	N.D.	N.D.	N.D.	—	—
Fireman *et al.* (1969)	Attenuated or killed 10 subjects, 5 virus, 5 placebo	Reduced by live vaccine only. Immediate sensitivity not affected. Microbial antigens (No interference by measles virus with function of transfer factor)	Unaffected	Decreased slightly 2 wks p.i.	Reduced	Cytological	20% foetal calf serum
Kadowaki *et al.* (1970) (letter only)	Natural measles	N.D.	Reduced but stated not by live vaccine	N.D.	N.D.	Cytological	?

Zweiman et al. (1971)	Live vaccine 8 subjects, 6 immune, 2 susceptible	PPD-5TU Decreased for 14 to 35 days	3 decreased 7 decreased (neutralized by autologous serum)	N.D.	1·0 μCi tritiated thymidine	Autologous serum
Kreth et al. (1974)	SSPE 15 patients 20 controls	6/14 completely negative all test agents. Others negative to some. But all sensitized by DNCB	Normal Normal	MLR Normal cell Cytotoxicity to carrier line. Strong by T-cells. Blocked by virus antigen not Ab	Tritiated thymidine uptake	15% foetal calf serum
McFarland (1974)	In mouse. Mouse-adapted strain	A suppression of the anti-hapten response. Heat-inactivated virus ineffective. Sindbis virus ineffective. Non-mouse-adapted virus ineffective. Effect declines from D10 up to D15 after inoculation. Suggests antigen-stimulated cells vulnerable to virus		—	—	—

reaction to phytohaemagglutinin was not. The killed vaccine reduced none of these reactions. Immediate hypersensitivity was not affected by vaccine although Fireman did note that following live vaccine eosinophil leukocytes disappeared from the circulation. It is also interesting to note that when measles vaccine and BCG vaccine were given simultaneously to two patients the time at which they became Mantoux positive was not less than normal, suggesting that ability to become sensitized and competence to show hypersensitivity are separate functions, one insensitive and the other sensitive to infection of the patient with measles. Tests of leukocytes *in vitro* confirmed that the response to phytohaemagglutinin was not significantly reduced by attenuated virus given to the patient, and a decreased *in vitro* response which was observed against purified protein derivative of tuberculin and against candida antigens returned to normal in about 4 to 8 weeks in different individuals (Fireman *et al.*, 1969).

The same techniques were found by Kadowaki and his colleagues (1970) to reveal inhibition of the lymphocyte response to phytohaemagglutinin during measles but not after vaccine, and Zweiman, using incorporation of thymidine into lymphocytes as an assay of responsiveness, reported that there was a depression of incorporation to be found in some patients to whom 1000 $TCID_{50}$ of Schwarz virus vaccine had been administered, when leukocytes were stimulated by phytohaemagglutinin and by purified protein derivative. Skin sensitivity was reduced for a period of 14 to 35 days after vaccination. The results were not related to the measles antibody titres of the patients, nor to the presence or absence of autologous serum in the *in vitro* tests (Zweiman *et al.*, 1971). When subacute sclerosing panencephalitis was examined as an example of measles infection *in vitro* cellular responses were active and normal although intradermal cutaneous reactions were diminished (Kreth *et al.*, 1974). Thus the persistence of the virus in the patient is not due to a generally deficient functioning of immuno-competent cells.

The cellular basis for diminished lymphocyte responses during infection with measles virus is unknown. One interesting set of experiments with a mouse-adapted strain of measles virus, testing the anti-hapten response in mice, has indicated that measles virus may specifically inhibit cells which take part in the control of autologous antibody production (McFarland, 1974). Such a mechanism might be important if certain clones of helper or suppressor cells were permanently affected by measles, for the resulting imbalance of immunological control might cause abnormally high or abnormally low responses to measles virus to persist even in the absence of further infection.

NON-SPECIFIC RESPONSES FOLLOWING
INFECTION *IN VITRO*

Infection of leucocytes *in vitro* also produces depression of certain immunological functions but experimental results are complicated in practice and in interpretation by the many factors and processes participating in the test system, some of which are certainly not measles virus or its components (see Table 11.2). A well-controlled test was carried out on 3 measles-immune and 3 non-immune subjects by Smithwick and Berkovich (1966) using purified protein derivative of tuberculin. Live measles virus, when added to lymphocyte cultures of all 6 patients, inhibited the cytological response to this antigen. Stimulation by phytohaemagglutinin was not reduced.

In contrast, a series of experiments by Zweiman has apparently shown that live measles vaccine *in vitro* and also inactivated measles vaccine both reduce the stimulating effect of phytohaemagglutinin and of purified protein derivative of tuberculin on sensitized lymphocytes as measured by the uptake either of tritiated thymidine or labelled amino acids (Zweiman, 1971, 1972; Zweiman and Miller, 1974), but there were complications due to the possible presence of other inhibiting substances in the virus preparation used; dialysed measles and the dialysate were both active in suppressing leukocyte reactivity. Thus the effect of virus was not proven but the effect of products of infection was. Recently a more clear cut demonstration has been given to the effect that the reduction of stimulation of lymphocytes by phytohaemagglutinin is produced by live measles virus and not by extraneous materials (Sullivan *et al.*, 1975). Here heat inactivation reduced the degree of inhibition and the residual inhibiting effect was indeed due to culture fluids which were inhibitory. A dose response was noted, best at a multiplicity of infection equal to 100, though measles did not kill all the cells, and virus had to be added within 8 h after exposure of the cells to phytohaemagglutinin in order to inhibit the stimulating effect of the mitogen. The experiments provide no data on whether the inhibitory action of measles virus is direct or mediated by products of infection, such as interferon, which is well produced by measles virus-infected leukocytes, but cell killing by virus was almost certainly not the cause of the reduced responses.

There are still technical sources of error to be considered. In a recent work Munyer and his colleagues have confirmed that immunization with live virus vaccine, this time containing measles, mumps and rubella viruses does significantly reduce the response to specific stimulation by candida antigen, but the apparent depression of non-specific reaction to phytohaemagglutinin or pokeweed mitogens disappeared when account was

TABLE 11.2

Effect of measles virus in vitro on lymphocyte stimulation

Reference	Measles virus	Subjects	Results of stimulus by PHA	Results of stimulus by PPD or OT	Assay	Medium
Smithwick and Berkovich (1966)	Live virulent	Measles 3 immune 3 susceptible	N.D.	Reduced (p. 0·01)	Cytological 1000 mono-nuclears counted	MEM foetal calf serum
Zweiman (1971)	Live attenuated 1×10^6 cells $1 \times 10^{5.7}$ TCD_{50}	8 immune 7 susceptible	Reduced $= 2\frac{1}{2} \times$ (neutralized by autologous serum)	(O.T.) Reduced $= 2\frac{1}{2} \times$	Tritiated thymidine at 5D.	E.M.E.M. foetal calf or with autologous serum
Zweiman (1972)	Live attenuated Inactive attenuated (autoclaved). 5×10^5 cells $10^{5.7}$ TCD_{50}	30 adults Not all tuberculin-positive	Reduced from 99% to 56%	(O.T.) Reduced from 98% to 46%	1·0 µCi tritiated thymidine at 3–5D also Autoradio-graphy.	E.M.E.M. 10% a-gamma globulinaemic foetal calf serum

Zweiman and Miller (1974)	Non-viable measles. Vaccine strains. Toxic effects excluded	19 experiments ? no of subjects	Reduced. Related to duration of incubation with virus	—	1·0 μCi tritiated thymidine at 3–5D also Autoradio-graphy	MEM 10% a-gamma globulinaemic foetal calf serum
Sullivan et al., (1975)	Live or inactivated	—	Reduced by live purified virus. (Non-specific effect of cell media virus-infected-cell media, etc., cleared up)	?		R.P.M.I. 1640 20% foetal calf serum + glutamine

taken of the non-specific changes in DNA turnover during virus infection (Munyer et al., 1975). This conclusion favours those reports which state that measles virus depresses immunogenic stimuli, not mitogenic stimuli.

In spite of one obvious complication, there is some measure of agreement to be seen in the preceding accounts that lymphocytes taken during measles or vaccine-measles, or infected with measles virus in vitro, are somewhat less responsive to mitogenic, and a good deal less responsive to antigenic, stimulation than are healthy lymphocytes. The principal complication is that lymphocytes from in vivo infection are not yet sensitized against measles virus; those used for in vitro infection are often taken from measles-immune subjects. Although care has been taken by a few authors to compare leukocytes from immune, with those from non-immune, individuals, the possibility of at least two contributing anti-genic stimuli occurring in measles-immune cell cultures must often be admitted. Other effects have been considered by different experimenters, namely, toxicity, cell-killing by the virus, interferon production and the presence of unrealized histo-compatibility antigens in part-purified measles virus suspension. These complexities have meanwhile prevented any elucidation of the mechanisms by which measles virus inhibits the different lymphocyte responses even in the fortunate circumstance that there is a good deal of agreement between the results of infection in vivo and added virus in vitro.

SPECIFIC RESPONSES TO MEASLES ANTIGEN

Just as measles or measles vaccine depressed delayed hypersensitivity to various antigens, so the reaction to measles itself may be absent. In testing an 8-year-old boy who had subacute sclerosing panencephalitis, the intracutaneous test against measles virus antigen was found to be absent, when the graft response and the reaction to the sensitizing effect of dinitrochlorobenzene were both present (Jabbour et al., 1969). The results suggest that in a measles-infected patient there is no loss of ability on the part of the cellular mechanisms to initiate sensitization, but there is a failure of some secondary event to respond fully to specific antigenic stimulation by the infecting measles virus.

There have been reports that measles virus often fails to excite a response in lymphocytes taken from people who are known to be immune to measles and whose lymphocytes may be presumed to be sensitized (Nelson et al., 1966; Lennon et al., 1967; Lennon and Isacson, 1967) but crude measles complement-fixing antigen was found recently not only to provide a consistently successful stimulus for sensitized lymphocytes but also to possess the particular characteristic of being more effective when poly AU was added to the reaction mixture (Chess et al., 1972; Graziano et al., 1975). That finding and others which have been obtained with various antigens including whole virus-infected cells as an antigenic stimulating agent appear in Table 11.3.

The variable quality of antigen makes it essential to find systems giving positive and negative reactions with different individual's lymphocytes in the same test. The following example illustrates the point well. Cultures of lymphocytes taken from patients with subacute sclerosing panencephalitis responded to virus-infected cells, but lymphocytes from measles-immune, normal individuals did not. Lymphocytes from the same patients did not, it was stated, respond to stimulation by purified active or inactivated measles virus. The response to measles virus-infected cells was significant and accompanied by similar responses to cytomegalovirus-infected cells only in those patients who had antibody to cytomegalovirus. The authors state that the two virus antigens which failed to stimulate-thymidine incorporation into leukocytes were quite active in suppressing macrophage migration. It should be noted that the measles virus in infected cells was derived from a case of subacute sclerosing panencephalitis (Thurman et al., 1973).

Sensitization of lymphocytes to an antigen can be assessed by a different type of test—the cytotoxicity of lymphocytes towards cells bearing the sensitizing antigen (Kreth et al., 1974). Cytotoxicity itself can be assayed by examining the degree of cell destruction or, more delicately, by measuring the release of chromium or some other tracer material from the target cells. It has been shown by these methods that the cells of measles-immune and non-immune subjects were active to some extent against control uninfected cells but equally so, whilst similar cells from measles immune patients demonstrated a remarkably destructive effect towards measles virus-infected cell lines, one of HeLa and one derived from W1 38 human diploid cells (Labowskie et al., 1974). It also seems from the results given in this paper that lymphocytes from measles-non-immune individuals were also distinctly more toxic to infected cells than to normal cells.

Recent work tends to confirm that both methods of assessing cell-mediated-immunity, either thymidine uptake into lymphocytes or cytotoxicity of lymphocytes to measles virus-infected cells, give comparable results, like the following.

Kreth and his colleagues, also studying the effect of lymphocytes on measles virus-infected cells, used Lu 106 cells carrying measles virus and chromium release from infected cells as a mark for immunological attack. They tested the lymphocytes of 10 patients in varying stages of

TABLE 11.3

Measles-specific lymphocyte responses

Reference	Subjects	Lymphocyte stimulation	Assay	Medium
Chen et al. (1969)	SSPE 1 case	Positive when skin test to measles lost	Graft response. DNCB sensitivity	—
Chess et al. (1972)	Normal	Weak, but amplified by poly AU	Tritiated thymidine	R.P.M.I. 1640 Autologous plasma.
Thurman et al. (1973)	SSPE ? No.	Cells persistently infected with SSPE measles or cytomegalo virus. Lymphocytic response, index 13·9, best at 72 h	Tritiated thymidine 1μCi (1·9 Ci/mmol) Not altered by removal of B. lymphocytes	R.P.M.I. 1640 in sero-negative plasma minus platelets. 2×10^5 lymphocytes $10^{4.0}$ to $2.5 \times 10^{4.0}$ target cells
Labowskie et al. (1974)	7 negative 12 positive history of measles 2 No history available	9/11 "positive" give greater Cr. release from measles infected cells than from non-infected cells.	Cytoxicity to persistently infected HeLa or W.I. 38. Chromium release	E.M.E.M. 20% autologous plasma 80–90% lymphocytes 5×10^6 lymphocytes 5×10^4 target cells

Utermohlen *et al.* (1974)	17 patients SLE 2 controls	Direct macrophage inhibition to measles virus only. Not to para-flu or rubella. Area 19·3% compared with 47·6%.	Area of migration	
Kreth *et al.* (1975)	SSPE 10 patients 20 controls	5×10^4 T lymphocytes at ratios 2/1 to 50/1. Spec. index up to 50% at 50/1 ratio. (Some sero-pos. individuals, non SSPE, strongly positive)	Cytotoxicity on Lu 106, persistent measles-infected. Chromium release. Clusters also seen	R.P.M.I. 1640, foetal calf serum, inactivated.
Graziano *et al.* (1975)	17 normal All positive history measles	Positive response 17/17, to measles CF antigen, inactivated β-propio-lactone. Suppressed by amantadine	Tritiated thymidine 2 μCi (1·9 Ci/mmol) 3×10^6 cells. Best at 7D PHA best at 4D	R.P.M.I. 1640 plus autologous plasma 15–20%

subacute sclerosing panencephalitis and cells of 10 adults and 10 children served as controls. At the highest ratios of lymphocyte to target cell, which was 50/1, the cytotoxic index given by cells from SSPE patients was raised above normal. The cytotoxic value of some adult and children's virus was quite high (Kreth et al., 1975).

Graziano and his colleagues, again using unpurified, complement-fixing antigens from measles virus-infected cells, on this occasion inactivated by β-propiolactone, tested the incorporation of tritiated thymidine into lymphocytes and showed that 17 samples of cells from 17 normal subjects all responded significantly to the measles antigen. The measles-specific reaction was best measured at the 7th day of incubation of the cells as opposed to the response to phytohaemagglutinin which was best done at the 4th day. The addition of poly AU again stimulated the response to measles virus antigen, but amantadine depressed it. The tests were carried out in autologous plasma (Graziano et al., 1975).

Patients' leukocytes have been subjected to the preceding tests in efforts to understand immunological defects which might account for chronic infection in subacute sclerosing panencephalitis or for the measles-specific serological changes present in multiple sclerosis. There are discordant results. For example, Klajman et al. (1973) record two cases of SSPE in which blast transformation by phytohaemagglutinin and by candida antigen was normal, whilst that to measles virus was much reduced and specific macrophage migration not different from normal. There was autologous plasma in the cultures. An opposite response to measles virus had already been recorded by Saunders et al., (1969), who used group AB human serum in the cell culture. In a summary of a series of 20 patients with SSPE, Moulias et al., (1971) assert that none showed a measles virus-specific response in the leukocyte migration inhibition test and in blast transformation, whereas control leukocytes from normal children were active in both tests. Similarly, Gerson and Haslam (1971) noted an absence of dermal sensitivity and absence of a leukocyte response to measles virus in 4 boys with the disease, but Lischner et al. (1972), commenting on Mizutani's demonstration, in a well-documented set of experiments in 1 patient, of a diminished dermal sensitivity to measles along with a diminished inhibition of macrophage migration (guinea-pig cells), give a list of contrasting experimental results that have been found by stimulating lymphocytes from patients with SSPE. We have referred to some of these. Amongst possible causes of discrepancy, Lischner et al. (1972) mention blocking antibody, since found by Sell et al. (1973).

COMMENT

The uncertainty and complexity of the tests makes it difficult to interpret these discrepant findings. We can refer only to the difficulties against which the results should be studied.

The first necessity is to produce a reliable set of tests. Dermal sensitivity to certain standard antigens should be fairly reliable, but in using measles antigens of mixed origin there is much less certainty. It should always be understood that when inhibition or reduction is found, there is no assurance that an immunological mechanism is at work. Reduction of non-immunological responses to codeine, for example, has been carefully recorded during measles and during scarlet fever (Pilcher, 1935).

A second difficulty lies in knowing whether lymphocyte responses always measure the same thing, even when like assay systems are in use. There is a suggestion that they may not. Purified measles antigens seem to induce blast transformation and stimulate the uptake of tritiated thymidine into leukocytes with difficulty, but induce inhibition of migration of sensitized cells readily; measles virus in cell fragments is active in lymphocyte transformation and intact measles virus-infected cells are excellent indicators of specific lymphocytic toxicity. It may be well to think of several pathways for the antigenic stimulus, just as there are two for the induction of complement activity, and to considerh ow to test these individually, when testing each patient's leukocytes.

We have already stressed the existence of many accessory factors in each test, such as cytotoxic and blocking antibodies in patients' serum and mutiple antigens in virus preparations. The ability of killed, as well as live, virus to produce interferon and other lymphokines through the action of its nucleic acid should be borne in mind in assessing whether synergistic or antagonistic mechanisms are likely to prevail over any specific response which is being sought.

The same applies to non-specific effects on the assay system. If, for example, the test antigen, such as whole virus, alters the measured response in non-sensitized cells, that is measles non-immune lymphocytes, the effect on sensitized cells should be measured against this in order to distinguish between absolute and relative changes in response. This has been emphasized by Munyer et al. (1975).

In spite of these difficulties it is possible to see some trends amongst the combined published tests of cell-mediated immunity in measles or in the presence of measles virus.

1. There is generally a depression of specific skin sensitivity and of in vitro lymphocyte reactions during measles and during infection by

some attenuated vaccines. Differences in response which have been noted between mitogenic and antigenic stimulation could be one of degree rather than kind although Munyer's results favour the alternative explanation. The antigenic response is more frequently and obviously depressed than the response to phytohaemagglutinin.

2. There are also depressed responses during *in vitro* infection with measles virus. It is difficult to say how similar results of the *in vivo* and *in vitro* infections really are. It is probable that most of the depressive action is due to the presence of live virus, but it is not all a direct effect. In culture, lymphocytes may be exposed to a greater concentration and variety of stimuli than individual lymphocytes *in vivo*, so increasing the difficulties of interpretation.

3. Different results have been found by different workers applying tests of lymphocyte function in the same disease, but in different patients. There is real difficulty in deciding whether the several co-operating and competing mechanisms in the various tests are altered by the disease or by the variety of conditions used. Here we shall ultimately be dependent on agreed standardization of method and on strict matching of controls, including matching for like immunological experience to the test agents. If consistent differences from normal then appear in any one disease they will have some validity; their meaning will have to await more understanding of the immunological processes at work.

4. Finally there is an overwhelming case for carrying out all virus-specific responses with single, purified antigens. Whole virus and even purified measles virus haemagglutinin will probably react with lymphocytes non-immunologically and even nucleocapsid may have one set of effects due to its antigenic properties and another set caused by its nucleic acid. Since the mode of presentation of the antigen, as cellular or "soluble" material has been found to influence the results of lymphocyte stimulation, once the effect of separate antigens is suspected, model, artificial, complexes will have to be devised having known sizes and defined compositions of antigen. In that way the influence of cellular antigens will be excluded from the tests.

Valdimarrson *et al.* (1975) and Nordal *et al.* (1975) advised against interpreting results of lymphocyte stimulation by measles virus as wholly due to immunological causes. These reservations appear in later sections.

12 | Immunology of measles virus—multiple sclerosis and auto-immune diseases

INTERPRETATION: (1) *Average titres of measles antibody are above normal in multiple sclerosis and in certain auto-immune conditions.* (2) *Cell-mediated responses seem to be less active than normal, including measles-specific responses.* (3) *Work on cell-mediated immunity is inadequate as yet.* (4) *There is no evidence that measles virus or immune responses to it have a causative role in these diseases.*

MULTIPLE SCLEROSIS

There is a set of relationships between measles virus and the immune system which is not yet understood and which is to be found in studies of multiple sclerosis and certain auto-immune diseases. In general terms, there is, in patients who have one of these diseases, a group tendency to display a slight hyperactivity of humoral immune responses and a hypoactivity of some of the cell-mediated immune reactions. This differs from the excessive antigenic stimulation which is induced by the large amount of measles antigen in the brain of patients who have subacute sclerosing panencephalitis and in whom antibody titres are increased and cellular responses undiminished (p. 165, 180, 184).

The apparent paradox in auto-immune disease and multiple sclerosis of hyperactivity in one part of the immune system and reduced activity in the other can only be referred to briefly because the inter-relating

TABLE 12.1

In vitro immunological reactivity of cells from multiple sclerosis patients

Reference	No. of patients	Mitogen	Response of leukocytes to stimulus by Antigen (non virus)	Measles or related virus antigen	Method of assay	Serum of medium
Fowler *et al.* 1966	6	—	CSF—increased	—	Cytology	Placental serum and autologous plasma.
Jensen, 1968	10	To PHA Decreased significantly	CSF—not increased	—	Thymidine incorporation	Autologous serum 33%.
Knowles and Saunders, 1970	8	—	—	Measles no different from controls.	Thymidine incorporation	10% pooled group AB serum
Day and Peterson, 1970	31	—	Highly significant increase to encephalitogenic protein.	Measles and RSV no different from controls, P.	Thymidine incorporation	Autologous serum 10% or "some" autologous.
Ciongoli *et al.* 1973; Dupont, 1974	18	—	PPD—no difference	Reduced response to measles, mumps, rubella, para-flu viruses.	Leukocyte migration in agarose.	Homologous serum (or MS serum. No difference in 4 tested).

Reference	No.					
Utermohlen and Zabriskie, 1973	15	—	Brain tissue antigen—no different from controls.	Reduced response to measles NOT to para-flu, rubella.	Leukocyte migration	Assumed autologous
Zabriskie, 1974	34	—	—	Reduced response to measles. NOT to rubella, mumps, para-flu.	Leukocyte migration	Transfer factor reversed the decrease in 8/10 tests.
Offner et al. 1974a	26	Decreased compared with controls.	Bovine encephalitic protein—no difference from controls.	No difference from controls. P.	Thymidine incorporation	—
Offner et al. 1974b	29	Decreased compared with controls.	To sheep encephalitic protein—no difference from controls.	Measles—decreased compared with controls (one antigen preparation only). P.	Myo-inositol incorporation	
Ciongoli et al. 1975	30	—	To candida or PPD—no difference from controls.	Virus 6/94 significant slight reduction compared with controls. P.	Leukocyte migration in agar	

P = purified antigen

functions of the relevant cell-mediated reactions have not yet been worked out. In particular, there are numerous variables in the different tests. The principal ill-defined factors include cell-derived lymphokines acting secondarily on target lymphocytes in the test culture, incompatibility between host antigens introduced with the virus and the test cells, mixed immunological and physiological effects of virus in the system, especially live virus, and anti-viral and anti-lymphocytic substances in the patient's serum. Table 12.1 contains data and references to works which show some findings in common in spite of the complexities involved in the experiments reported.

Serology in multiple sclerosis

The original observations of Adams and Imagawa (1962) were that multiple sclerosis patients have a slightly higher average titre of circulating measles virus-specific antibody than do other patients or healthy people. Given that samples are large enough and factors such as age, sex, race, country and place of residence are equated, Adams' original claim can be substantiated (for recent reviews see Brody *et al.*, 1972; Cathala and Brown, 1972; Fraser, 1975). The difference in mean titre between the normal groups and multiple sclerosis patients is not large, seldom more than two-fold, and surveys which present discrepant findings tend to be too small. Surveys which tend to implicate other viruses in the same way as measles are again small and have mostly used one type of test, the complement fixation test. The significance of this has not been investigated, but, for many viruses, it is not the most sensitive indicator available for detecting antibody.

Technique has some bearing on results with measles virus. Titres of anti-haemolysin in multiple sclerosis patients are more significantly different from normal than titres of anti-haemagglutinin, when the same sets of sera are compared (Salmi *et al.*, 1973). On the other hand, when measles virus-specific antibody to nucleocapsid is examined in spinal fluid, it is disproportionately present in that site three times more commonly than other measles virus-specific antibodies (Vandvik and Degre, 1975) which are themselves more often disproportionately present than antibodies specific for other viruses (for a review see Cathala and Brown, 1972). Disproportion here refers to the convention of measuring the ratio of antibody titre in cerebro-spinal fluid to the titre of the same antibody in serum. The ratio in health is usually low 1:300 to 1:500 for antibodies to different viruses and enhanced ratios not only indicate a disproportionate amount of antibody in cerebrospinal fluid, but they are also taken to indicate that some synthesis of antibody is going on within the central nervous system. Ratios of other serum constituents, such as albumin, or a different virus antibody are always checked simultaneously as being within normal ranges.

If the disproportionate ratio of measles antibody in CSF to that in serum is confined to measles virus, it is of some importance to the pathogenesis of the disease multiple sclerosis and also to the biology of measles virus. There might, conceivably, be latent infection in the brain. We feel that the results have been a little biased in favour of measles virus by lack of interest in other membrane viruses for which ratios have not been studied. Measles virus-specific antibodies are more commonly found in CSF of multiple sclerosis patients than other antibodies are.

Haire's results (Haire et al., 1973b) suggest that equally great and equally frequent disproportion may be found with antibodies to herpes simplex and rubella viruses, sometimes in the same patient in whom the measles antibody ratio is similarly disturbed. CSF/serum antibody ratios of other enveloped RNA viruses need to be studied carefully in order to decide whether the intra-cerebral distribution of virus antibody in multiple sclerosis patients is unique to measles or whether it merely indicates that when antibody producing cells occur in the brain, the individual patient's general extra-cerebral capacity to manufacture antibodies of different virus-specificity ensures that measles antibody is the commonest.

One indubitable fact is that much of the immunoglobulin of spinal fluid from multiple sclerosis patients is less heterogeneous, as judged by electrophoretic measurements, than the gamma globulin in normal serum and spinal fluids. That is, it is oligoclonal (Link, 1967; Kolar et al., 1970; Link et al., 1973; Norrby et al., 1974) and some of it is measles virus-specific (Panelius and Salmi, 1973). A recent report claims that it may be further restricted by belonging to one subclass of IgG, namely IgG_1 (Palmer et al., 1976). These findings make it almost certain that measles virus-specific IgG and perhaps some IgG of other restricted specificities is being synthesized in the brain. Measurement of the dynamics of IgG synthesis in CSF had already suggested this to Tourtellotte (1975).

A final measles-specific difference between multiple sclerosis patients and others is marked by a circulating IgM called MS-IgM which appears in just over one-third of patients and which binds to an acetone-sensitive and phospholipase C-sensitive receptor at the surface of measles virus-infected cells in tissue culture (Millar et al., 1971; Haire et al., 1973a, b; Fraser et al. 1974). Its significance is unknown and it is not found in other diseases. It has not been found in cerebro-spinal fluid and in serum it seems to fluctuate in titre, or appear and disappear during the disease, though its relationship to clinical state has yet to be determined (Haire

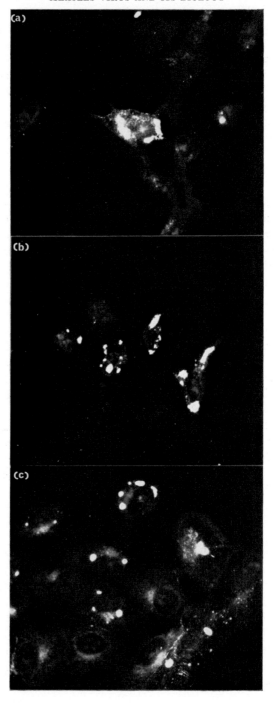

et al., unpublished data). Recently Gharpure *et al.* (unpublished data) (Fig. 12.1) have detected a globulin of similar specificity in guinea-pig measles-immune sera*. It is possible that MS-IgM indicates a continuing low-grade stimulation by measles virus or by some antigenically related organism.

There are those who suspect from the preceding observations that there is a possible aetiological, but unexplained, relationship between measles and multiple sclerosis. The disproportion between CSF/serum ratios of measles antibody, oligoclonal antibody in cerebrospinal fluid and circulating measles specific IgM all provide tests which can be applied to individuals, as opposed to groups of patients. Some specific activity of measles virus (or a cross-reacting virus) or some persisting abnormality of the measles virus-specific immune response must be present in multiple sclerosis patients. Observations which show whether these are primary events or are phenomena secondary to disease, will be awaited with interest.

Lymphocyte responses in multiple sclerosis

Records of normal cellular responses to measles antigen are in a much less clearcut state. In addition to the variables which we have referred to (p. 180), measles virus antigens have proved to be rather too weak as inducers of *in vitro* cell responses, and also of dermal hypersensitivity, to show clearcut differences between patients.

With the foregoing facts in mind, the emerging state of the relationship between multiple sclerosis and the cellular immunology of measles may be gathered from the data in Table 12.1. There is probably a general non-specific depression of lymphocyte responses in multiple sclerosis as seen by the response to mitogenic stimuli (Jensen, 1968; Offner *et al.*,

* Both MS-IgM and the guinea-pig globulin have since been identified as specific for measles virus haemolysin (Iidem, unpublished data).

FIG. 12.1   Staining of an acetone-sensitive and phospholipase C-sensitive measles virus-specific membrane antigen by a guinea-pig anti-measles serum. (a) Indirect fluorescent antibody staining by guinea-pig measles-immune serum, of UNFIXED, infected HEp_2 cells. Treated with phospholipase C, 1 unit/ml., for 5 min at 37°C. Central cell shows intracellular aggregates of measles antigen and staining of the cell membrane. Uninfected cells, with no inclusions, do not show membrane fluorescence. (b) From the same culture treated 20 min with phospholipase C. No membrane staining of inclusion-bearing cells. (c) From the same culture. Non-treated cells FIXED in acetone 10 min at room temperature. No membrane staining of inclusion bearing cells. The staining of inclusions is specifically anti-nucleocapsid; the staining of membrane is anti-haemolysin.

1974a, b). At present nothing points to the cause of this; that is whether the defect is innate in the immunocompetent cell or acquired from external factors such as metabolites or proteins acting upon the cells. In contrast, the response to antigenic stimuli such as candida albicans or purified protein derivative of tuberculin is not much different in multiple sclerosis from that in normal patients' lymphocytes. This seems to apply also to measles virus antigen, when thymidine incorporation is used as a measure of cell stimulation (Knowles and Saunders, 1970; Dau and Peterson, 1970; Offner et al., 1974a). However, the measurement of leukocyte migration and the incorporation of myo-inositol into stimulated cells, which are both primary membrane functions, have both shown a depressed response to measles virus antigens, as compared to other antigens and opposite from the reaction induced by measles virus antigens in normal persons (Ciongoli et al., 1973; Dupont, 1974; Utermohlen and Zabriskie, 1973; Zabriskie, 1975; Offner et al., 1974b). The only deficiencies amongst these results refer to specificity; are they wholly measles specific or do other enveloped RNA viruses expose the same depression of function when acting on lymphocytes from multiple sclerosis patients? If measles virus is going to be related specifically to multiple sclerosis, it is necessary to decide whether measles and other paramyxoviruses, which are physically alike and act on cell membranes in similar ways, induce different responses in lymphocytes derived from patients who have multiple sclerosis but the same reactions in lymphocytes from patients who do not have that condition.

The fundamental problem for the immunologist is to discover how the homeostasis of the immune system can be upset so as to cause simultaneously excess activity of B cells forming antibody against measles virus and depressed measles-specific activity of other lymphocyte functions. A virus-induced inhibition of suppressor T cells is a likely possibility.

We have not considered cutaneous reactions to injected measles virus; there are not enough reported results (Sever and Kurtzke, 1969).

SYSTEMIC LUPUS ERYTHEMATOSIS AND AUTO-IMMUNE STATES

We refer in this group to diseases in which the presence of organ-specific or tissue-specific antibodies is well known.

Raised titres of antibody against measles virus have been noticed in systemic lupus erythematosis and in Reiter's disease and, occasionally, in some rheumatoid or other collagen diseases (Phillips and Christian, 1969, 1970).

There is one notable difference from the immunological findings in multiple sclerosis, in that raised titres of antibody specific for parainfluenza viruses and perhaps for some others, not all of which are RNA enveloped viruses, are also found. Both measles and parainfluenza antibodies were estimated by the anti-haemagglutinin test, but Hollinger and his colleagues found that the complement-fixation test gave confirmatory results, revealing the greatest difference to be between viruses in patients with systemic lupus erythematosis, less amongst other connective tissue diseases and least in normal control subjects. Measles virus showed clearly significant differences in mean antibody titres between those groups; mumps virus, parainfluenza viruses, rubella and reoviruses showed the same tendencies, but with lesser degrees of statistical significance (Hollinger et al., 1971). It may also be of some consequence to immune function that 21 out of 26 sera tested from patients with lupus erythematosis had anti-leukocytic antibodies. These have also been noticed in multiple sclerosis (Stjernholm et al., 1970; van den Noort and Stjernholm, 1971) in which they are related to clinical activity; autocytotoxic factors have also been reported (Kuwert and Bertrams, 1972). To these findings Lucas added the important observation that slightly raised titres of measles virus-specific antibodies are more common in patients who have got auto-immune antibody to one of several different tissue antigens (Lucas et al., 1972). Thus the immune response to measles virus is implicated in two diseases or groups of diseases which have one or other aspect of an auto-immune disease. The basis of the relationship is unknown.

Antibodies to measles virus have attracted less notice in connective tissue diseases than in multiple sclerosis and even less attention has been paid to measles-specific cell-mediated reactions. One recent report is worth considering because it concerns the effect of measles virus on direct macrophage migration using leukocytes taken from patients with systemic lupus erythematosis (Utermohlen et al., 1974), and using the same antigens that had been applied in tests on cells from multiple sclerosis patients (Utermohlen and Zabriskie, 1973). When parainfluenza virus, rubella virus and measles virus were each used to test the same samples of leukocytes from patients with systemic lupus erythematosis, the inhibition by measles virus was very significantly less than the inhibition by the other two viruses, relative to the inhibition produced on lymphocytes from patients who did not have systemic lupus erythematosis. As in multiple sclerosis measles-specific cellular immunity seems to be depressed in lupus erythematosis.

COMMENT

Because of ignorance about the mechanisms of cell-mediated immunity many assumptions have had to be made and difficulties ignored in drawing conclusions about the effect of measles virus on lymphocyte responses. Their validity is also affected by the fact that results are few and samples small. If the depression of measles-specific cell-mediated immunity in multiple sclerosis patients as compared with others exists, and if it is comparable in degree to the slight excess in mean titre of measles-specific antibody, then large test populations or accumulated results in many smaller groups will be required to show it conclusively.

Unlike antibody titration, testing lymphocyte responses may stimulate antagonistic processes. We have already seen in Chapter 11 that live measles virus *in vitro* depresses immune cellular responses. The depressive non-immunological effect and any stimulated antigenic response may well cancel each other out. It will, therefore, be essential to use single, purified antigens when testing immune responses.

Very few surveys have employed immunological matching in comparing test and control populations. It may be safe to assume that in samples of middle-aged patients and control subjects that experience of measles virus will be the same in both groups. It is not safe to assume that this holds for all viruses, when comparing the effects of measles virus with others. Contradictory results amongst different surveys could be so explained if the number of individuals tested is small. Antibody tests for the various reagents would serve to sort out the like and the unlike before applying tests of cellular function to groups of people.

Enough has been done to justify further exploration of the findings which suggest that changes in cell-mediated immunity will be found in patients who have multiple sclerosis or auto-immune disease. The present state of affairs is very similar to the early claims and contradictions which followed Adams' and Imagawa's paper about measles antibody in multiple sclerosis. Only further work will tell.

Efforts will have to be made to disentangle the non-immunological effects of measles virus on lymphocytes from its immunological effects, by avoiding the use of living virus and also by employing antigens of the virus which do not of themselves find non-immunological receptors on cell membranes. When such crude information has been obtained from studies on properly matched groups of control and test patients the various stages by which the antigenic stimulus travels inwards and the response outwards will have to be found and examined, before the site of action of measles virus can be discovered. This achievement is some way off.

13

Immunology of measles virus—immunization; immunodeficient states

INTERPRETATION: (1) *Live attenuated measles virus is a most effective vaccine.* (2) *Abnormal sensitization, which occurs during vaccination, and abnormal pathological conditions which occur during immunodeficiency or immunodepression, point to variation of the host rather than variation in the virus as the main cause.* (3) *The effects are characteristic of measles virus rather than other viruses in similar circumstances, implying that some highly specific virus–cell relationships are involved, probably immune functions.*

IMMUNIZATION

We need not consider the practical and administrative problems which have been overcome in order to establish that measles vaccine is a safe and effective prophylactic agent, but there are side-effects in the individual and problems for the community which are connected with attributes of measles virus. The extent to which measles immunization has been applied and the gradual acceptance of live vaccine in place of inactivated vaccine during recent years may be learned from various reviews (Zhdanov, 1961; Krugman, 1971; Linneman, 1973).

Effects of vaccination

The conversion rate from susceptible to immune status is measured by the antibody response. Even when the least sensitive measure, the complement-fixation test, is used, attenuated vaccine shows a 97% to 100% successful conversion rate (Katz et al., 1962; Hilleman et al., 1968). A small number of infectious doses of vaccine can be effective (Schwarz, 1964) and larger doses are not necessarily more effective (Popov et al., 1973). The circulating antibody titre is related to the degree of immunity (Karelitz and Markham, 1962) but absence of circulating antibody need not mean susceptibility to measles, for a-gamma-globulinaemic patients can be immune to challenge for at least 3 years after vaccination (Katz et al., 1962). There seems to be no natural transmission from person to person of the live vaccine and in a closed community the mean titre of antibody falls with time (Brown et al., 1969; Black and Rosen, 1962).

Killed measles vaccine is less effective than attenuated, for the anti-body titre attained after vaccination falls gradually, even after a booster inoculation; for example, Rauh and Schmidt (1965) noted that the proportion of children whose serum had circulating measles-specific antibody fell from 98% to 69% over a 12-month period and, of 145 children from the group who were later exposed to measles, no less than 43·2% developed it.

It is well known that infants who are immunized early in the first year of life respond relatively poorly to vaccine. This applies also to measles and to live vaccine. It is on record that two children, vaccinated at 2 and 3 months of age and who had maternal antibody to measles at that time, did not show a rise of titre after being vaccinated against measles and still had no measles virus-specific antibody $1\frac{1}{2}$ years and 2 years later (Halonen et al., 1965). On the other hand passively admin-istered anti-measles gamma globulin does not reduce the conversion rate that is produced by live measles vaccine (Linneman, 1973).

The effect of previous immunity upon vaccination is known. Children who have no measles-specific antibody develop measles-specific immuno-globulin M in response to immunization, whilst those who have cir-culating antibody do not readily do so (Linneman, 1973).

Complications of vaccination

Complications of measles following vaccination have been recorded, including deaths from encephalitis, but very few of that small number have been shown to be caused by the vaccine and, of these, several have been known and others suspected, to occur in patients who were immuno-logically incompetent for various reasons.

Cases of subacute sclerosing panencephalitis have also been related in time to measles immunization; one of the first to be recognized occurred within 3 weeks of vaccination and the patient survived for 17 months (Schneck *et al.*, 1967). Others have been reported (Schneck, 1968; Payne and Baublis, 1971; Parker *et al.*, 1970; Gerson and Haslam, 1971), but there has been no rise in the incidence of SSPE as a consequence of world-wide use of attenuated measles strains (Krugman, 1971). This, Krugman suggests, means that attenuated and wild measles have equal chances of causing this chronic intra-cerebral infection. The incidence of acute encephalitis has fallen in proportion to the decline in measles notifications (Fig. 13.1).

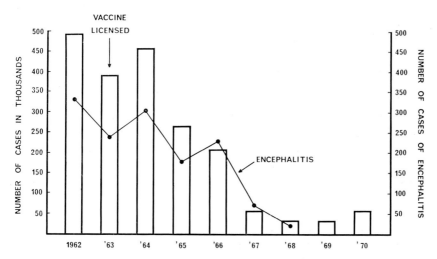

FIG. 13.1   Reported cases of measles and measles encephalitis in the U.S.A., 1962–1970. (From Krugman (1971), with permission.) A decrease in the number of cases of measles and a proportionate decrease in the prevalence of measles encephalitis, after vaccine was licensed, suggests that encephalitis is not a frequent complication of vaccination.

Live measles vaccines have the capacity to produce mild systemic disease and partial immunity to these effects does result from prior administration of killed vaccine. Moreover the average antibody titre produced by live vaccine in persons already vaccinated by the killed material may exceed the specific response in non-immunized vaccinees and no untoward results need follow (Karelitz *et al.*, 1965). Effectiveness, nevertheless, is greater after live vaccine alone, 93% as compared with 85% (Report, Measles Vaccine Committee 1971). However, soon after the administration of killed vaccine was adopted as a policy designed to avoid unwanted side-effects of live, attenuated vaccine, and particularly when alum-precipitated material had been employed, a series of reports appeared indicating that a small proportion of people who had been immunized some time before did develop unusual reactions after exposure to measles or to measles vaccine. When natural infection followed immunization with killed vaccine, a few patients reacted badly. Rashes were atypical in appearance and in distribution, being marked by petechaie and oedema of hands and feet; there were often respiratory symptoms, sometimes high fever and often polymorphonuclear leukocytosis instead of the expected leukopoenia (Rauh and Schmidt, 1965; Fulginiti et al., 1967; Nader et al., 1968; Gokiert and Beamish, 1970; McLean et al., 1970). Radiological examination showed infiltration of the lungs (Buser and Montagnon, 1970). Similar abnormal signs and symptoms followed the administration of live vaccine in some of those who had already been immunized with killed measles virus with the addition of fairly well-marked, Arthus-like reactions at the site of inoculation (Buser, 1967; Harris, 1967; McNair Scott and Bonanno, 1967). Occasionally similar symptoms have been found also, but in milder form, when live vaccine succeeded primary immunization with attenuated, not killed, measles virus. Nothing is known of the mechanisms which produce this, but the ratio of complement-fixing antibody to anti-measles-haemagglutinin in some such cases is said to be abnormally high (Thiry et al., 1969).

Little investigation has been done into the pathogenesis of this interesting form of hypersensitivity, but undoubtedly the alum-precipitated material is sensitizing in rabbits (Bonin et al., 1971) in which tests relate the hypersensitivity to the animal protein in which virus was grown, rather than to virus antigen. Nevertheless, the reaction to natural measles implies that sensitivity can also be induced to an antigen in the virus itself. Whether there is a cross-reacting sensitivity between virus and host tissue is unknown and this is a point of some importance to the possible induction of autoimmunity by measles in natural circumstances. Quantitative effects of excess preformed specific antibody should also be investigated.

The phenomenon of post-vaccinal hypersensitivity to subsequent infection is not confined to measles virus, but has been observed after immunization against respiratory syncytial virus, trachoma and mycoplasmas, as noted by Rød et al., (1970) in describing their series of atypical measles patients. All these antigens contain much lipid. The non-virus sensitizing component is derived from tissue culture and is protein (Aprile et al., 1969) not lipid, as claimed (Hennessen and Mauler, 1967), but this does not explain hypersensitivity to natural infection.

Comment

Two practical questions remain unanswered at present. In the event of killed measles vaccine being preferred in some circumstances, such as the need to overcome difficulties of transport and storage in remote places, in which killed vaccine is more stable than live, which of the virus antigens is most capable of inducing protective immunity? Secondly, is the sensitivity induced by some killed vaccines due to virus antigen or to cellular material that can be eliminated from the vaccine? We have recently been impressed by the fact that an antibody prepared against highly purified measles virus grown in "Vero" monkey kidney cells, cross-reacted strongly with a human tissue culture line HEp_2. The significance of this is not known. The cross-reacting material was not removed by absorption with calf tissue, although calf serum was common to the two cell cultures, but only after absorption of the serum with HEp_2 cells (Gharpure, unpublished data). A cross-reacting host antigen in the virus envelope seems to be a likely explanation.

IMMUNODEFICIENCY AND IMMUNOSUPPRESSION

Three patients suffering from giant-cell pneumonia, without a rash, were found by Enders and his colleagues to have a generalized infection by measles virus. The virus was recovered but only on cultures of human kidney cells, from lungs, liver and spleen of all 3 patients (Enders *et al.*, 1959).

Since then, giant-cell pneumonia, accompanied usually by microscopic evidence of widespread invasion by measles virus, has been recognized as being particularly associated with leukaemia and similar states in which the functioning of the immune system is severely depressed as well as being found in patients who have had a recognizable deficiency in immunological function (Mawhinney *et al.*, 1971; Haram and Jacobsen, 1973). Sometimes pneumonia is the first sign that measles virus is present. In leukaemia patients excretion of virus is prolonged and the antibody response delayed but it is probable that observations like this refer to patients receiving immuno-suppressive drugs (Mitus *et al.*, 1959). It seems unlikely that any particular strain of measles virus, or any particular characteristic of the virus, can account for the ability to take advantage of the inhibited immunological functions of the patient other than the close affinity which measles virus always has with cells of the lymphoid system (see pathogenesis, p. 6). Even attenuated virus has been found in association with fatal systemic measles in immunosuppressed and immunodeficient patients (Mitus *et al.*, 1962, 1965; Mawhinney *et al.*,

1971; Mihatsch *et al.*, 1972). The immunosuppressive effect of measles virus itself might be considered an extra burden in this context, yet immunodepression similar to that produced by measles virus is known to accompany influenza (Bloomfield and Mateer, 1919) and poliomyelitis (Berkovich and Starr, 1966), and generalized systemic lesions due to these two viruses do not ensue. The lymphotropism of measles virus seems to be the significant aetiological characteristic in this respect.

One of the reports mentioned shows that a child who had sufficient immuno-competence to respond well to other sorts of virus vaccine by developing specific antibody, succumbed to infection when live attentuated measles was given (Mawhinney *et al.*, 1971). The only other connection with immune deficiency seems to be with low circulating levels of non-specific immunoglobulin A (Eibl *et al.*, 1973).

The absence of rash in many of the patients is interesting and must represent a defect of cell-mediated immunity, for, again, patients who were hypogammaglobulinaemic have been reported to develop typical measles rashes (Good and Zak, 1956).

Many descriptions of individual patients contain points of interest on the immunology and pathogenesis of measles. One notable example of subacute sclerosing panencephalitis has been described in which there was also pneumonia and no antibody to measles virus was present; the patient had leukaemia (Breitfeld *et al.*, 1973). This at least indicates that the pathogenesis of encephalitis in subacute sclerosing panencephalitis can be due to the cytopathic effect of the virus and not to incidental serological mechanisms.

Immunosuppression in a child being treated for a nephrotic syndrome had apparently a similar effect to immunodeficiency, when measles supervened. Death followed in 19 days with no formation of antibody, although diphtheria antitoxin arising from immunization 3 years before was detectable in the patient's serum (Meadow *et al.*, 1969). This event was unfortunate in view of the occasional beneficial effects of measles in nephrosis, but the general cytotoxic effect of treatment does not allow us to say which immunological function is related to the fatal progress of the measles virus.

The effects of immunosuppression have been studied in experimental measles encephalitis. A strain of virus, which after intrathalamic inoculation into rhesus monkeys grew little, produced few signs of encephalitis and caused much microscopic inflammation in the brain, was found in cyclophosphamide-treated monkeys to produce much virus antigen in the brain and to cause severe encephalitis, but the inflammatory response was now practically absent. Here at least the inflammatory response was beneficial (Albrecht *et al.*, 1972). However, the same strain of virus and another mouse-adapted strain which were partly lethal to weanling mice, when inoculated intra-cerebrally, were no more lethal during cyclophosphamide treatment of the mice than they were in untreated mice. The specific antibody response was suppressed for the 18 days duration of the experiment (Griffin et al., 1974). One explanation for the difference between the two species may be that infection in weanling mice is abortive (p. 136) and the negative results of immuno-suppression therefore not valid. The importance of this issue, the pathogenic or the therapeutic effects of the immune response during measles encephalitis, has been well expressed by Koprowski (1962).

Apart from the relationship between immunological function and pathogenesis, there are a few instances of lack of permanent acquired immunity to measles virus. One well annotated family history of specific susceptibility to measles relates how typical attacks, diagnosed by consultant paediatricians, recurred every 2 years in a boy of 8 years of age (Thamdrup, 1952). There was a history of similar events in other members of the family. Serum proteins were normal, as judged at that time, and the boy developed good antibody in response to injections of diphtheria and tetanus toxoids.

A short description, of different relevance, concerns deficiency in immunoglobulin A. In nine different forms of this deficiency, iso-agglutinins, antibodies to mumps virus, to Bordetella pertussis, to diphtheria and tetanus toxins were all within the normal range. But antibodies specific for measles virus were absent or below the normal range in all nine patients (Eibl et al., 1973). The significance of this remains unknown.

COMMENT

The various immune responses to measles and the relationships to disease which have been considered point rather to variation in the patient than to mutation in the virus. However, genetic variants apart, the physiological behaviour of the virus—the quantity produced, its location, complete or incomplete replication, temporary or persistent infection, the number and amount of virus antigens at the cell surface—all must have profound effects on the nature of the antigenic stimulus received by the host. Yet, only some patients given killed, alum-precipitated vaccine became abnormally hypersensitive; only the leukaemic and immunologically suppressed or immuno-deficient respond to infection by developing systemic progressive measles with giant cell pneumonia, only patients in the group of auto-immune diseases or those with multiple sclerosis have slightly overactive antibody responses to measles along with partially depressed cell-mediated responses. All that we can say at present is that most other viruses, even those with similar structure and natural history, do not do this. There must be some close relationship between measles and the immune system which remains to be discovered and explained.

References

Ackerman, P. H. and Black, F. L. (1961). A radiobiological analogy between measles virus and temperate phages. *Proc. natl. Acad. Sci. U.S.A.* **47**, 213–220.

Ackerman, A. B. and Suringa, D. W. R. (1971). Multinucleate epidermal cells in measles. *Arch. Derm.* **103**, 180–184.

Adams, J. M. and Imagawa, D. T. (1957a). Immunological relationship between measles and distemper viruses. *Proc. Soc. exp. Biol. Med.* **96**, 240–244.

Adams, J. M. and Imagawa, D. T. (1957b). The relationship of canine distemper to human respiratory disease. *Ped. Clin. North Am.* **4**, 193–201.

Adams, J. M. and Imagawa, D. T. (1962). Measles antibodies in multiple sclerosis. *Proc. Soc. exp. Biol. Med.* **111**, 562–566.

Adams, J. M., Imagawa, D. T., Yashimori, M. and Huntington, R. W. (1956). Giant cell pneumonia. Clinicopathologic and experimental studies. *Pediatrics*, **18**, 888–898.

Adams, J. M., Imagawa, D. T., Wright, S. W. and Tarjan, G. (1959). Measles immunization with live avian distemper virus. *Virology*, **7**, 351–353.

Adels, B. R., Gajdusek, C., Gibbs, C. J., Albrecht, P. and Rogers, N. G. (1968). Attempts to transmit subacute sclerosing panencephalitis and isolate a measles related agent, with a study of the immune response in patients and experimental animals. *Neurology*, **18**, 30–51.

Ahmed, A., Strong, D. M., Sell, K. W., Thurman, G. B., Knudsen, R. C., Wistar, R. and Grace, W. R. (1974). Demonstration of a blocking factor in the plasma and spinal fluid of patients with subacute sclerosing panencephalitis. I. Partial characterisation. *J. exp. Med.* **139**, 902–924.

Albrecht, P. and Schumacher, H. P. (1971). Neurotropic properties of measles virus in hamsters and mice. *J. infect. Dis.* **124**, 86–93.

Albrecht, P. and Schumacher, H. P. (1972). Markers for measles virus: I Physical properties. *Arch. ges. Virusforsch.* **36**, 23–35.

Albrecht, P., Shabo, A. L., Burns, G. R. and Tauroso, N. M. (1972). Experimental measles encephalitis in normal and cyclophosphamide-treated rhesus monkeys. *J. infect. Dis.* **126**, 154–161.

Anderson, G. D. and Atherton, J. G. (1964). Effect of Actinomycin D on measles virus growth and interferon production. *Nature, Lond.* **203**, 670–671.

Anderson, J. F. and Goldberger, J. (1911). Experimental measles in the monkey: A preliminary note. *Public Hlth. Reports (U.S.)*, **26**, 847–848.

Aprile, M. A., Kadar, D. and Wilson, S. (1969). Sensitization of guinea-pigs to measles vaccines. *Clin. exp. Immunol.* **5**, 531–539.

Arakawa, S. (1948). Experimental study on measles virus. *Jap. J. med. Sci. Biol.* **1**, 12–14.

Arakawa, S. (1949). Experimental study on measles virus. *Jap. J. exp. Med.* **20**, 351–366.

Arakawa, S. (1964). Recent advances in measles virology. *Ergeb. Mikrobiol. Immunol.* **38**, 1–38.

Arita, M. and Matumoto, M. (1968). Plaque formation and initial virus-cell interaction of measles virus. *Jap. J. Microbiol.* **12**, 51–61.

Askerov, V. F. and Voroshilova, M. K. (1968). Investigation of the effect of measles infection and vaccination with live measles virus vaccine upon the status of humoral immunity to poliomyelitis in children. *Vopr. Virusol.* **1**, 42–47.

Avakova, A. N. and Rapoport, R. I. (1971). The effect of the measles virus on chromosomal apparatus of diploid cell strains. *Tsitologiia*, **13**, 830–836.

Babbot, F. L. and Gordon, J. E. (1954). Modern measles. *Amer. J. Med. Sci.* **228**, 334–361.

Barbanti-Brodano, G., Possati, L. and La Placa, M. (1971). Inactivation of polykaryocytogenic and hemolytic activities of Sendai virus by phospholipase B (Lysolecithinase). *J. Virol.* **8**, 796–800.

Baringer, J. R. and Griffith, J. F. (1970). Experimental measles virus encephalitis. A light, phase, fluorescence and electron microscopic study. *Lab. Invest.* **23**, 335–346.

Barron, A. L., Milgrom, F., Karzon, D. T. and Witebsky, E. (1963). Demonstration of human measles antibody by mixed agglutination. *J. Immunol.* **90**, 908–913.

Barry, D. W., Sullivan, J. L., Lucas, S. L., Dunlap, R. C. and Albrecht, P. (1976). Acute and chronic infection of human lymphoblastoid cell lines with measles virus. *J. Immunol.*, **116**, 89–98.

Bartlett, M. S. (1957). Measles periodicity and community size. *Jl. R. stat. Soc.* **120**, 48–60.

Bather, R., Furesz, J., Fanok, A. G., Gill. S. D. and Yarosh, W. (1973). Long-term infection of diploid African green monkey brain cells by Schwarz measles vaccine virus. *J. gen. Virol.* **20**, 401–405.

Bech, V. (1958). Studies on measles virus in monkey kidney tissue cultures. 2. Development of cytopathic changes and identification of the cultivated agents by complement fixation tests. *Acta path. microbiol. scand.* **42**, 86–96.

Bech, V. (1960a). Titers of complement fixing measles antibodies in human sera collected from one to five years after illness. *Acta path. microbiol. scand.* **50**, 81–88.

Bech, V. (1960b). Relationship between complement fixing antibodies against measles virus and canine distemper virus. *Acta path. microbiol. scand.* **50**, 331–334.

Bech, V. (1960c). Titers of complement fixing measles antibodies in sera collected at random from healthy persons in Copenhagen. *Acta path. microbiol. scand.* **50**, 322–330.

Bech, V. (1962). Measles epidemics in Greenland. *Amer. J. dis. Child.* **103**, 252–253.

Bech, V. and von Magnus, P. (1958). Studies on measles virus in monkey kidney tissue cultures. 1. Isolation of virus from five patients with measles. *Acta. path. microbiol. scand.* **42**, 75–85.

Bellanti, J. A., Sanga, R. L., Klutinis, B., Brandt, B. and Artenstein, M. S. (1969). Antibody responses in serum and nasal secretions of children immunized with inactivated and attenuated measles virus vaccines. *New Engl. J. Med.* **280**, 628–633.

Benjamin, B. and Ward, S. M. (1932). Leukocytic response to measles. *Amer. J. dis. Child.* **44**, 921–963.

Ben-Porat, T., Lonis, B. and Kaplan, A. S. (1975). Further characterization of a population of defective interfering pseudorabies virions. *Virology*, **65**, 179–186.

Bentzon, J. W. (1953). The effect of certain infectious diseases on tuberculin allergy. *Tubercle*, **34**, 34–41.

Bergholz, C. M., Kiley, M. P. and Payne, F. E. (1975). Isolation and characterization of temperature-sensitive mutants of measles virus. *J. Virol.* **16**, 192–202.

Berkovich, S. and Starr, S. (1966). Effects of live type 1 polio virus vaccine and other viruses on the tuberculin test. *New Engl. J. Med.* **274**, 67–72.

Black, F. L. (1959a). Serological epidemiology in measles. *Yale J. Biol. Med.* **32**, 44–50.

Black, F. L. (1959b). Growth and stability of measles virus. *Virology*, **7**, 184–192.

Black, F. L. (1966). Measles endemicity in insular populations: Critical community size and its evolutionary implication. *J. theoret. Biol.* **11**, 207–211.

Black, F. L. and Rosen, L. (1962). Patterns of measles antibodies in residents of Tahiti and their stability in the absence of re-exposure. *J. Immunol.* **88**, 725–731.

Black, F. L. and Sheridan, S. R. (1967). Blood leukocyte response to live measles vaccine. *Amer. J. dis. Child.* **113**, 301–304.

Black, F. L., Reissig, M. and Melnick, J. L. (1956). Propagation of measles virus in a strain of human epidermoid cancer cells (Hep-2). *Proc. Soc. exp. Biol. Med.* **93**, 107–108.

Black, F. L., Reissig, M. and Melnick, J. L. (1959). Measles virus. *Adv. vir. Res.* **6**, 205–227.

Blair, C. D. and Duesberg, P. H. (1970). Myxovirus ribonucleic acids. *Ann. Rev. Microbiol.* **24**, 539–574.

Blair, C. D. and Robinson, W. S. (1968). Replication of Sendai virus. I. Comparison of the viral RNA and virus-specific RNA synthesis with Newcastle disease virus. *Virology*, **35**, 537–549.

Blake, F. G. and Trask, J. D. (1921a). Studies on measles. I. Susceptibility of monkeys to the virus of measles. *J. exp. Med.* **33**, 385–412.

Blake, F. G. and Trask, J. D. (1921b). Studies on measles. II. Symptomatology and pathology in monkeys experimentally infected. *J. exp. Med.* **33**, 413–422.

Bloomfield, A. L. and Mateer, J. G. (1919). Changes in skin sensitiveness to tuberculin during epidemic influenza. *John Hopkins Hosp. Bull.*, **342**, 238–239.

Blumberg, R. W. and Cassady, H. A. (1947). Effect of measles on the nephrotic syndrome. *Amer. J. dis. Child.* **73**, 151–166.

Bogaert, L van. (1945). Une leuco-encéphalite sclèrosante subaiguë. *J. neurol. Neurosurg. Psychiat.* **8**, 101–120.

Bonin, O., Schmidt, I., Schmidt, K., Enders-Ruckle, G. and Berkes, I. (1971). Arthus phenomenon-like skin reaction and antibody pattern in rabbits immunized with various myxovirus fractions. *Arch. ges. Virusforsch.* **33**, 211–224.

Boyle, D. B. and Atherton, J. G. (1971). Immunochemical detection of measles virus in BSC-1 cells. *Arch. ges. Virusforsch.* **35**, 299–302.

Breese, Jr. S. S., De Boer, C. J. (1973). Ferritin-tagged antibody cross-reactions among rinderpest, canine distemper and measles viruses. *J. gen. Virol.* **20**, 121–125.

Breitfeld, V., Hashida, Y., Sherman, F. E., Odagiri, K. and Yunis, E. J. (1973). Fatal measles infection in children with leukaemia. *Lab. Invest.* **28**, 279–291.

Brincker, J. A. H. (1936). The control of measles. *Lancet i*, 103–107.

Brody, J. A. and Detels, R. (1970). Subacute sclerosing panencephalitis: A zoonosis following aberrant measles. *Lancet ii*, 500–501.

Brody, J. A., Sever, J. L., Edgar, A. and McNew, J. (1972). Measles antibody titres of multiple sclerosis patients and their siblings. *Neurology*, **22**, 492–499.

Brown, L. V. (1957). Pathogenicity for rabbit kidney cell cultures of certain agents derived from "normal" monkey tissue. 1. Isolation and propagation. *Amer. J. Hyg.* **65**, 189–209.

Brown, P., Gajdusek, D. C. and Tsai, T. (1969). Persistence of measles antibody in the absence of circulatory natural virus five years after immunization of an isolated virgin population with Edmonston B. vaccine. *Amer. J. Epidem.* **90**, 514–518.

Burnstein, T. and Byington, D. P. (1968). On the isolation of measles virus from infected brain tissue. *Neurology*, **18**, 162–164.

Burnstein, T., Jensen, J. H. and Waksman, B. H. (1964). The development of a neurotropic strain of measles virus in hamsters and mice. *J. infect. Dis.* **114**, 265–272.

Burnstein, T., Jacobsen, L. B., Zeman, W. and Chen, Tsu T. (1974). Persistent infection of BSC-1 cells by defective measles virus derived from subacute sclerosing panencephalitis. *Infection and Immunity*, **10**, 1378–1382.

Buser, F. (1967). Side reaction to measles vaccination suggesting the Arthus phenomenon. *New. Engl. J. Med.* **277**, 250–251.

Buser, F. and Montagnon, B. (1970). Severe illness in children exposed to natural measles after prior vaccination against the disease. *Scand. J. infect. Dis.* **2**, 157–160.

Bussell, R. H., Waters, D. J., Seals, M. K. and Robinson, W. S. (1974). Measles, canine distemper and respiratory syncytial virions and nucleocapsids. A comparative study of their structure, polypeptide and nucleic acid composition. *Med. Microbiol. Immunol.* **160**, 105–124.

Buynak, E. B., Peck, H. M., Creamer, A. A., Goldner, H. and Hilleman, M. R. (1962). Differentiation of virulent from avirulent measles strains. *Amer. J. dis. Child.* **103**, 460–473.

Byington, D. P. and Burnstein, T. (1973). Measles encephalitis produced in suckling rats. *Exp. Mol. Path.* **19**, 36–43.

Byington, D. P. and Johnson, K. P. (1972). Experimental subacute sclerosing panencephalitis in the hamster: correlation of age with chronic inclusion-cell encephalitis. *J. infect. Dis.* **126**, 18–26.

Byington, D. P. and Johnson, K. P. (1975). Subacute sclerosing panencephalitis virus in immunosuppressed adult hamsters. *Lab. Invest.* **32**, 91–97.

Carlström, G. (1957). Neutralisation of canine-distemper virus by serum of patients convalescent from measles. *Lancet, ii,* 344.

Carlström, G. (1958). Comparative studies on measles and distemper viruses in suckling mice. *Arch. ges. Virusforsch.* **8**, 527–538.

Carlström, G. (1962). Relation of measles to other viruses. *Amer. J. dis. Child.* **103**, 287–291.

Carter, C., Black, F. L. and Schluederberg, A. (1973a). Nucleus associated RNA in measles virus-infected cells. *Biochem. Biophys. Res. Commun.* **54**, 411–416.

Carter, C., Schluederberg, A. and Black, F. L. (1973b). Viral RNA synthesis in measles virus-infected cells. *Virology,* **53**, 379–383.

Cascardo, M. and Karzon, D. T. (1965). Measles virus giant cell inducing factor (Fusion Factor). *Virology,* **26**, 311–325.

Castro, A. E., Burnstein, T. and Byington, D. P. (1972). Properties in cell culture of a hamster brain-adapted subacute sclerosing panencephalitis virus. *J. gen. Virol.* **16**, 413–417.

Cathala, F. and Brown, P. (1972). The possible viral aetiology of disseminated sclerosis. *J. clin. Path.* 25 Suppl. **6**, 141–151.

Cello, R. M., Moulton, J. C. and McFarland, S. (1959). The occurrence of inclusion bodies in the circulating neutrophils of dogs with canine distemper. *Cornell. Veterinarian.* **49**, 127–144.

Cernescu, C., Cajal, N. and Draganescu, N. (1972). The neuropathogenicity of some measles virus strains for the newborn hamster. *Rev. Roum. Virol.* **9**, 183–189.

Cernescu, C., Cajal, N., Cepleanu, M. and Sorodoc, Y. (1973). Approaches to attenuation of an autochthonous measles virus strain. *Rev. Roum. Virol.* **10**, 117–126.

210   REFERENCES

Chen, T. T., Watanabe, I., Zeman, W. and Mealey, J. (1969). Subacute sclerosing panencephalitis: propagation of measles virus from brain biopsy in tissue culture. *Science*, **163**, 1193–1194.

Chen, C., Compans, R. W. and Choppin, P. W. (1971). Parainfluenza virus surface projections: glycoproteins with haemagglutinin and neuraminidase activities. *J. gen. Virol.* **11**, 53–58.

Chess, L., Levy, C., Schmukler, M., Smith, K. and Mardiney, M. R. (1972). The effect of synthetic polynucleotides on immunologically induced tritiated thymidine incorporation. *Transplantation*, **14**, 748–755.

Chiang, T.-P. (1966). Propagation of measles virus in developing chick embryos. *Biken Journal.* **9**, 131–133.

Chiarini, A. and Norrby, E. (1970). Separation and characterisation of products of two measles virus variants. *Arch. ges. Virusforsch.* **29**, 205–214.

Choppin, P. W. and Compans, R. W. (1970). Phenotypic mixing of envelope proteins of the parainfluenza virus SV5 and vesicular stomatitis virus. *J. Virol.* **5**, 609–616.

Choppin, P. W., Klenk, H. D., Compans, R. W. and Caliguiri, L. A. (1971). The parainfluenza virus SV5 and its relationship to the cell membrane. *In:* "Perspectives in Virology, VII". (M. Pollard, ed.) pp. 127–156. Academic Press, London and New York.

Choppin, P. W., Scheid, A. and Mountcastle, W. E. (1975). Paramyxoviruses, membranes and persistent infections. *Neurology*, **25**, 494.

Chu, Liang-wei and Morgan, H. R. (1950). Studies of the haemolysis of red blood cells by mumps virus. 1. The development of mumps virus haemolysin and its inactivation by certain physical and chemical agents. 2. The relationships of haemagglutination, virus elution and haemolysis. *J. exp. Med.* **91**, 393–416.

Ciongoli, A. K., Platz, P., Dupont, B., Svejgaard, A., Fog, T. and Jersild, C. (1973). Lack of antigen response to myxoviruses in multiple sclerosis. *Lancet ii*, 1147.

Ciongoli, A. K., Lisak, R. P., Zweiman, B., Koprowski, H. and Waters, D. (1975). *In vitro* cellular responsiveness in multiple sclerosis patients to a viral isolate from multiple sclerosis brain tissue and to other antigens. *Neurology*, **25**, 891–893.

Closs, O., Haukenes, G., Gjone, E. and Blomhoff, J. P. (1971). Raised antibody titres in chronic liver disease. *Lancet ii*, 1202–1203.

Closs, O., Haukenes, G., Blomhoff, J. P. and Gjone, E. (1973). High titres of antibodies against rubella and morbilli virus in patients with chronic hepatitis. *Scand. J. Gastroent.* **8**, 523–528.

Collins, B. S. and Bratt, M. A. (1973). Separation of the messenger RNAs of Newcastle disease virus by gel electrophoresis. *Proc. nat. Acad. Sci. U.S.A.* **70**, 2544–2548.

Compans, R. W. and Choppin, P. W. (1968). The nucleic acid of the parainfluenza virus SV5. *Virology*, **35**, 289–296.

Connolly, J. (1972). Subacute sclerosing panencephalitis. *J. clin. Path.* 25 *Suppl.* **6**, 73–77.

Connolly, J. H., Allen, I. V., Hurwitz, L. J. and Millar, J. H. D. (1967). Measles-virus antibody and antigen in subacute sclerosing panencephalitis. *Lancet i*, 542–544.

Connolly, J. H., Haire, M. and Hadden, D. S. M. (1971). Measles immuno-globulins in subacute sclerosing panencephalitis. *B.M.J.* **1**, 23–25.

Cutchins, E. C. (1962). A comparison of the haemagglutination-inhibition, neutralisation and complement fixation tests in the assay of antibody to measles. *J. Immunol.* **88**, 788–795.

Dahlberg, J. E. and Simon, E. H. (1969). Physical and genetic studies of Newcastle disease virus: Evidence for multiploid particles. *Virology*, **38**, 666–678.

Dalldorf, G., Douglass, M. and Robinson, H. E. (1938). Canine distemper in the rhesus monkey (Macaca Mulatta). *J. exp. Med.* **67**, 323–332.

Dau, P. C. and Peterson, R. D. A. (1970). Transformation of lymphocytes from patients with multiple sclerosis. *Arch. Neurol.* **23**, 32–40.

Dawson, J. R. Jr. (1933). Cellular inclusions in cerebral lesions of lethargic encephalitis. *Amer. J. Path.*, **9**, 7–15.

Dayan, A. D. and Stokes, M. I. (1971). Immunofluorescent detection of measles virus antigens in cerebrospinal-fluid cells in subacute sclerosing panencephalitis. *Lancet i*, 891–892.

Dekking, F. (1965). Study on the persistence of measles antibodies. *Arch. ges. Virusforsch.* **16**, 208–209.

Dekking, F. and McCarthy, K. (1956). Propagation of measles virus in human carcinoma cells. *Proc. Soc. exp. Biol. Med.* **93**, 1–2.

De Maeyer, E. (1960). Plaque formation by measles virus. *Virology*, **11**, 634–638.

De Meio, J. L. (1962). Measles virus hemolysin. *Virology*, **16**, 342–344.

De Meio, J. L. and Gower, T. A. (1961). Hemagglutination by measles virus. *Virology*, **13**, 367–368.

Denton, J. (1925). The pathology of fatal measles. *Amer. J. med. Sci.* **169**, 531–543.

Desmyter, J. J. L., Melnick, and Rawls, W. E. (1968). Defectiveness of inter-feron production and of rubella virus interference in a line of African green monkey kidney cells (Vero). *J. Virol.* **2**, 955–961.

Detels, R., Brody, J. A., McNew, J. and Edgar, A. H. (1973). Further epide-miological studies of subacute sclerosing panencephalitis. *Lancet ii*, 11–14.

Draganescu, N., Cernescu, C., Nereantiu, F., Girjabu, E., Cepleanu, M. and Cajal, N. (1972). Investigations on experimental encephalitis induced by the etiologic agent of subacute sclerosing panencephalitis (SSPE). *Rev. Roum. Virol.* **9**, 191–196.

Draganescu, N., Dumitrescu, S. M., Cernescu, C., Girjabu, E., Cajal, N. and Pacuraru, E. (1973). Electron optical investigations on the morphological alterations produced by subacute sclerosing panencephalitis (SSPE) virus in *Cercopithecus aethiops* (R_6 CA) kidney cells. *Rev. Roum. Virol.* **10**, 33–37.

Dubois-Dalcq, M. and Barbosa, L. H. (1973). Immunoperoxidase stain of measles antigen in tissue culture. *J. Virol.* **12**, 909–918.

Dubois-Dalcq, M., Worthington, K., Gutenson, O. and Barbosa, L. H. (1975). Immunoperoxidase labelling of subacute sclerosing panencephalitis virus in hamster acute encephalitis. *Lab. Invest.* **32,** 518–526.

Duc-Nguyen, H. and Rosenblum, F. N. (1967). Immunoelectron microscopy of the morphogenesis of mumps virus. *J. Virol.* **1,** 415–429.

Dupont, B. (1974). Histocompatibility linked immune responsiveness in autoimmune diseases and possible implications for immunostimulatory therapy with special reference to multiple sclerosis. *In*: "Multiple Sclerosis Research". (A. N. Davison *et al.*, eds), pp. 291–314. H.M.S.O. Lond.

East, J. and Kingsbury, D. W. (1971). Mumps virus replication in chick embryo lung cells: properties of RNA species in virions and infected cells. *J. Virol.* **8,** 161–173.

Ehrengut, W. (1965). Measles encephalitis: Age disposition and vaccination. *Arch. ges. Virusforsch.* **16,** 311–314.

Ehrnst, A. C. (1975). Characterization of measles virus-specific cytotoxic antibodies by use of a chronically infected cell line. *J. Immunol.* **114,** 1077–1082.

Eibl, M., Götz, M., Kreppler, P. and Zimprich, H. (1973). Observations on the combination of IgA-deficiency and extremely low antibody titers against measles. *Z. Immunitatsfor. exp. Imm.* **145,** 183–190.

Enders, J. F. (1956). Observations on certain viruses causing exanthematous diseases in man. *Amer. J. med. Sci.* **231,** 622–637.

Enders, J. F. (1962). Measles virus. *Amer. J. dis. Child.* **103,** 282–287.

Enders, J. F. and Peebles, T. C. (1954). Propagation in tissue cultures of cytopathogenic agents from patients with measles. *Proc. Soc. exp. Biol. Med.* **86,** 277–286.

Enders, J. F., Peebles, T. C., McCarthy, K., Milovanovic, M., Mitus, A. and Holloway, A. (1957). Measles virus: A summary of experiments concerned with isolation, properties and behaviour. *Amer. J. Public Hlth.* **47,** 275–282.

Enders, J. F., McCarthy, K., Mitus, Anna and Cheatham, W. J. (1959). Isolation of measles virus at autopsy in cases of giant-cell pneumonia without rash. *New Engl. J. Med.* **261,** 875–881.

Enders, J. F., Katz, S. L. and Grogan, Elizabeth. (1962). Markers for Edmonston virus. *Amer. J. dis. Child.* **103,** 473–474.

Enders-Ruckle, G. (1963). Untersuchungen zum mechanismus der Masernimmunität. *Zbl. Bakt. 1 Orig.* **191,** 217–236.

Enders-Ruckle, G. (1965). Methods of determining immunity, duration and character of immunity resulting from measles. *Arch. ges. Virusforsch.* **16,** 182–207.

Fenner, F. (1948). The pathogenesis of the acute exanthems. An interpretation based on experimental investigations with mouse-pox (infectious ectromelia of mice). *Lancet ii,* 915–920.

Fiala, M., Payne, J. E., Berne, T. V., Moore, T. C., Henle, W., Montgomerie, J. Z., Chatterjee, S. N. and Guze, L. B. (1975). Epidemiology of cytomegalovirus infection after transplantation and immunosuppression. *J. infect. Dis.* **132**, 421–433.

Finch, J. T. and Gibbs, A. J. (1968). Observations on the structure of the nucleocapsids of some paramyxoviruses. *J. gen. Virol.*, **6**, 141–150.

Finkeldey, W. (1931). Über Riesenzellbefunde in den Gaumenmandeln, zugleich ein Beitrag zur Histopathologie der Mandelveränderungen im Maserninkubationsstadium. *Virchow's Arch.* **281**, 323–329.

Fireman, P., Friday, G. and Kumate, J. (1969). Effect of measles vaccine on immunologic responsiveness. *Pediatrics*, **43**, 264–272.

Florman, A. L. and Agatston, H. J. (1962). Keratoconjunctivitis as a diagnostic aid in measles. *J.A.M.A.* **179**, 568–570.

Follett, E. A. C., Pringle, C. R., Pennington, T. H. and Shirodaria, P. (1975). Measles virus development in enucleate cells. Abstracts: International Congress in Virology, Madrid, p. 180.

Follett, E. A. C., Pringle, C. R., Pennington, T. H. and Shirodaria, P. (1976). Events following the infection of enucleate cells with measles virus. *J. gen. Virol.*, **32**, 163–175.

Fowler, I., Morris, C. E. and Whitley, T. (1966). Lymphocyte transformation in multiple sclerosis induced by cerebrospinal fluid. *New Engl. J. Med.* **275**, 1041–1044.

Frankel, J. W. and West, M. K. (1958). Cultivation of measles virus in stable line of human amnion cells. *Proc. Soc. exp. Biol. Med.* **97**, 741–742.

Frankel, J. W., Burnstein, T., Denney, J. T. and West, M. K. (1957). Studies on isolates of measles viruses. *Bact. Proc.* 76.

Frankel, J. W., Burnstein, T. and West, M. K. (1958a). Propagation of measles virus in tissue cultures of dog kidney cells. *Fed. Proc.* **17**, item 1995.

Frankel, J. W., Cooke, H., Deviney, J. T., Warner, M. and West, M. K. (1958b). Serological response of guinea-pigs to inactivated measles. *Bact. Proc.* 58.

Fraser, K. B. (1959). Features of the MEL × NWS systems in influenza A virus. I. The effect of age of the mouse on the intracerebral growth of the MEL strain of influenza A virus. *Virology*, **9**, 168–177.

Fraser, K. B. (1975). Population serology. *In* "Multiple Sclerosis Research" (A. N. Davison *et al.*, eds), H.M.S.O. Lond. pp. 53–68.

Fraser, K. B., Shirodaria, P. V. and Haire, M. (1974). Enzyme treatment of unfixed cell cultures as a means of determining different specificities of immunoglobulin-M in multiple sclerosis and measles. *Med. Microbiol. Immunol.* **160**, 221–230.

Fraser, K. B., Gharpure, M., Shirodaria, P. V., Armstrong, M. A. and Dermott, E. (1977). Virus membrane protein in measles virus-infected cells. *In* "Negative strand viruses and the host cell." (R. D. Barry and B. W. J. Mahy eds.), Academic Press. London and New York. In press.

Freeman, J. M., Magoffin, R. L., Lennette, E. H. and Herndon, R. M. (1967). Additional evidence of the relation between subacute inclusion body encephalitis and measles virus. *Lancet ii*, 129–131.

Fulginiti, V. A., Eller, J. J., Downie, A. W. and Kempe, C. H. (1967). Altered reactivity to measles virus. Atypical measles in children previously immunized with inactivated measles virus vaccines. *J. Amer. med. Ass.* **202**, 1075–1080.

Funahashi, S. and Kitawaki, T. (1963). Studies on measles virus hemagglutination. *Biken J.* **6**, 73–96.

Gerson, K. L. and Haslam, R. H. A. (1971). Subtle immunologic abnormalities in four boys with subacute sclerosing panencephalitis. *New Engl. J. Med.* **285**, 78–81.

Geddes, T. D. and Gregory, W. J. (1974). Transferrin immunoglobulins and prognosis in measles in the tropics. *Trop. geogr. Med.*, **26**, 79–83.

Gibbs, F. A., Gibbs, E. L., Carpenter, P. R. and Spies, H. W. (1959). Electroencephalographic abnormality in "uncomplicated" childhood diseases. *J.A.M.A.* **171**, 1050–1055.

Gibson, P. E. and Bell, T. M. (1972). Persistent infection of measles virus in mouse brain cell cultures infected *in vivo*. *Arch. ges. Virusforsch.* **37**, 45–53.

Gillespie, J. H. and Karzon, D. T. (1960). A study of the relationship between canine distemper and measles in the dog. *Proc. Soc. exp. Biol. N.Y.* **105**, 547–551.

Gokiert, J. G. and Beamish, W. E. (1970). Altered reactivity to measles virus in previously vaccinated children. *Can. Med. Assoc. J.* **103**, 724–727.

Good, R. A. and Zak, S. J. (1956). Disturbances in gamma globulin synthesis as "experiments of nature". *Pediatrics*, **18**, 109–149.

Goodall, E. W. (1925). Measles with an "Illness of Infection". *Clin. J.* **54**, 69–71.

Gordon, H. and Knighton, H. T. (1941). Experimental measles. The lymphoid tissues of animals inoculated with the virus of human measles. *Amer. J. Path.* **17**, 165–176.

Goret, P., Mornet, P., Gilbert, Y. and Pilet, C. (1957). Immunité crosiése entre la maladie de Carré et la peste bovine. *C. R. Acad. Sci.* **245**, 2564–2566.

Goret, P., Fontaine, J., Mackowiack, C. and Pilet, C. (1959). Neutralisation du virus de la maladie de Carré par le serum contre la peste bovine. *C. R. Soc. Biol.* **248**, 2143–2144.

Gould, E. (1974). Variants of measles virus. *Med. Microbiol. Immunol.* **160**, 211–219.

Gould, E. A. and Linton, P. E. (1975). The production of a temperature-sensitive persistent measles virus infection. *J. gen. Virol.* **28**, 21–28.

Gould, E. A., Cosby, S. L. and Shirodaria, P. V. (1976). Salt-dependent haemagglutinating measles virus in SSPE. *J. gen. Virol.* **33**, 139–142.

Grado, C. and Ohlbaun, A. (1973). The effect of rifampicin, Actinomycin D and Mitomycin C on poliovirus and FMDV replication. *J. gen. Virol.* **21**, 297–303.

Granoff, A. (1962). Heterozygosis and phenotypic mixing with Newcastle disease virus. *Cold Spring Harb. Symp. quant. Biol.* **27**, 319–326.

Graziano, K. D., Ruckdeschel, J. C. and Mardiney, M. R. (1975). Cell-associated immunity to measles (Rubeola). The demonstration of *in vitro* lymphocyte tritiated thymidine incorporation in response to measles complement fixation antigen. *Cell. Immunol.* **15**, 347–359.

Greenberg, M., Pellitteri, O. and Eisenstein, D. T. (1955). Measles encephalitis. 1. Prophylactic effect of gamma globulin. *J. Pediat.* **46**, 642–647.

Greenham, I. W., Peacock, D. B., Hill, T. J., Brownell, B. and Schutt, W. H. (1974). The isolation of SSPE measles virus in newborn mice. *Arch. ges. Virusforsch.* **44**, 109–120.

Greenwood, M. (1935). "Epidemics and Crowd-diseases". pp. 180–196. Williams and Norgate. London.

Gresser, I. and Chany, C. (1963). Isolation of measles virus from the washed leucocytic fraction of blood. *Proc. Soc. exp. Biol. Med.* **113**, 695–698.

Griffin, D. E., Mullinix, J., Narayan, O. and Johnson, R. T. (1974). Age dependence of viral expression: Comparative pathogenesis of two rodent-adapted strains of measles virus in mice. *Infect. Immun.* **9**, 690–695.

Grist, N. R. (1950). The pathogenesis of measles: Review of the literature and discussion of the problem. *Glasgow Med. J.* **31**, 431–441.

Haire, M., Fraser, K. B. and Millar, J. H. D. (1973a). Virus-specific immunoglobulins in multiple sclerosis. *Clin. exp. Immunol.* **14**, 409–416.

Haire, M., Fraser, K. B. and Millar, J. H. D. (1973b). Measles and other virus-specific immunoglobulins in multiple sclerosis. *B.M.J.* **2**, 612–615.

Hall, W. W. (1974). Biochemical studies on measles virus. Ph.D. Thesis. Queen's University, Belfast.

Hall, W. W. and Martin, S. J. (1973). Purification and characterization of measles virus. *J. gen. Virol.* **19**, 175–188.

Hall, W. W. and Martin, S. J. (1974a). The biochemical and biological characteristics of the surface components of measles virus. *J. gen. Virol.* **22**, 363–374.

Hall, W. W. and Martin, S. J. (1974b). Structure and function relationships of the envelope of measles virus. *Med. Microbiol. Immunol.* **160**, 143–154.

Hall, W. W. and Martin, S. J. (1975). The structural proteins of measles virus, In: "Negative Strand Viruses" (B. W. J. Mahy and R. D. Barry, eds). pp. 89–103. Academic Press, London and New York.

Hall, W. W. and ter Meulen, V. (1976). RNA homology between subacute sclerosing panencephalitis and measles viruses. *Nature, Lond.* **246**, 474–477.

Hall, W. W., Martin, S. J. and Gould, E. (1974). Defective interfering particles produced during the replication of measles virus. *Med. Microbiol. Immunol.* **160**, 155–164.

Halonen, P., Forssell, P. and Backman, A. (1965). A follow-up study of immunity in children vaccinated with live attenuated measles virus vaccine. *Arch. ges. Virusforsch.* **16**, 268–272.

Hamilton, R., Barbosa, L. and Dubois, M. (1973). Subacute sclerosing panencephalitis measles virus: study of biological markers. *J. Virol.* **12**, 632–642.
Häntzschel, K. (1940). Masern rezidive. *Münch. med. Wschr.* **87**, 1178–1180.
Haram, K. and Jacobsen, K. (1973). Measles and its relationship to giant cell
pneumonia (Hecht pneumonia). *Acta Path.* **81**, 761–769.
Harnden, D. G. (1964). Cytogenetic studies on patients with virus infections
and subjects vaccinated against yellow fever. *Amer. J. Human. Genet.*
16, 204–213.
Harris, H. (1974). "Nucleus and Cytoplasm." pp. 137. Clarendon Press.
Harris, M. J. (1967). Reactions to live measles vaccine in children previously
given killed measles vaccine. *Med. J. Aust.* **54**, 989–990.
Harter, D. H. and Tellez-Nagel, I. (1968). Attempts to isolate SSPE agent in
cell culture. *Neurology*, **18**, 133–137.
Haspel, M. V. and Rapp, F. (1975). Measles virus: An unwanted variant
causing hydrocephalus. *Science*, **187**, 450–451.
Haspel, M. V., Duff, R. and Rapp, F. (1975a). Isolation and preliminary
characterization of temperature-sensitive mutants of measles virus. *J.
Virol.* **16**, 1000–1009.
Haspel, M. V., Duff, R. and Rapp, F. (1975b). Experimental measles ence-
phalitis: a genetic analysis. *Infect. Immun.* **12**, 785-790.
Haspel, M. V., Knight, P. R., Duff, R. C. and Rapp, F. (1973). Activation of a
latent measles virus infection in hamster cells. *J. Virol.* **12**, 690–695.
Hathaway, B. M. (1935). Generalised dissemination of giant cells in lymphoid
tissue in prodromal stage of measles. *Arch. Path.* **19**, 879–824.
Hayashi, K., Niwa, A., Rosenthal, J. and Notkins, A. L. (1973–74). Detection of virus-induced membrane and cytoplasmic antigens. Comparison
of the ^{125}I-labeled anti-viral antibody binding technique with immunofluorescence. *Intervirology*, **2**, 48–51.
Havens, W. P. and Marck, R. E. (1946). The leukocytic response of patients
with experimentally-induced infectious hepatitis. *Amer. J. med. Sci.*
212, 129–138.
Hecht, V. (1910). Die Riesenzellenpneumonie im Kindesalter, eine historisch-
experimentelle Studie. *Beitr. path. Anat.* **48**, 263–310.
Hennessen, W. and Mauler, R. (1967). Hypersensitization by measles virus.
Lancet i, 902.
Herndon, R. M. and Rubinstein, L. J. (1968). Light and electron microscopy
observation on the development of viral particles in the inclusions of
Dawson's encephalitis (subacute sclerosing panencephalitis). *Neurology*,
18, 8–20.
Herndon, R. M., Rena-Descalzi, L., Griffin, D. E. and Coyle, P. K. (1975). Age
dependence of viral expression. Electron microscopic and immunoperoxi-
dase studies of measles virus replication in mice. *Lab. Invest.* **33**, 544–553.
Hilleman, M. R., Buynak, E. B., Weibel, R. E., Stokes, J., Whitman, J. E. and
Leagus, M. B. (1968). Development and evaluation of the Movaten
measles virus vaccine. *J. Amer. med. Assoc.* **206**, 587–590.

Hodes, D. S., Schnitzer, T. J., Kalica, A. R., Camargo, E. and Chanock, R. M. (1975). Inhibition of respiratory syncytial, parainfluenza 3 and measles viruses by 2-deoxy-D glucose. *Virology,* **63,** 201–208.

Holland, J. J. and Villarreal, L. P. (1974). Persistent non-cytocidal vesicular stomatitis virus infections mediated by defective T particles that suppress virion transcriptase. *Proc. nat. Acad. Sci. U.S.A.* **71,** 2956–2960.

Hollinger, F. B., Sharp, J. T., Lidsky, M. D. and Rawls, W. E. (1971). Antibodies to viral antigens in systemic lupus erythematosis. *Arth. Rheum. N.Y.* **14,** 1–11.

Homma, M. (1975). Host-induced modification of Sendai virus. *In:* "Negative Strand Viruses" (B. W. J. Mahy and R. D. Barry, eds), Vol. II, pp. 685–697. Academic Press, London and New York.

Horta-Barbosa, L. Fuccillo, D. A., Sever, J. L. and Zeman, W. (1969a). Subacute sclerosing panencephalitis: Isolation of measles virus from a brain biopsy. *Nature. Lond.* **221,** 974.

Horta-Barbosa, L., Fuccillo, D. A., London, W. T., Jabbour, J. T., Zeman, W. and Sever, J. L. (1969b). Isolation of measles virus from brain cell cultures of two patients with subacute sclerosing panencephalitis. *Proc. Soc. exp. Biol. Med.* **132,** 272–277.

Horta-Barbosa, L., Fuccillo, D. A., Hamilton, R., Traub, R., Ley, A. and Sever, J. L. (1970). Some characteristics of SSPE measles virus. *Proc. Soc. exp. Biol. Med.* **134,** 17–21.

Horta-Barbosa, L., Hamilton, R., Wittig, B., Fuccillo, D. A. and Sever, J. L. (1971). Subacute sclerosing panencephalitis: Isolation of suppressed measles virus from lymph node biopsies. *Science,* **173,** 840–841.

Horvath, L. B. (1973). Effect of some factors on the biological activity of measles virus. *Acta microbiol. Acad. Sci. hung.* **20,** 83–88.

Hosaka, Y. and Shimizu, Y. K. (1972). Artificial assembly of enveloped particles of HVJ (Sendai Virus). I. Assembly of hemolytic and fusion factors from solubilized envelopes by Nonidet P. 40. *Virology,* **49,** 627–639.

Hosaka, Y., Kitano, H. and Ikeguchi, S. (1966). Studies on the pleomorphism of HVJ virions. *Virology,* **29,** 205–221.

Hövel, H. (1971). Einflus der Medium—Osmolarität auf die Virus-Plaque-Bilding im permanenten Zellkulturen. *Zbl. Bakt. 1. Abt. Orig.* **217,** 419–430.

Howe, C., Morgan, C., de Vaux St. Syr. C., Hsu, K. C. and Rose, H. M. (1967). Morphogenesis of type 2 parainfluenza virus examined by light and electron microscopy. *J. Virol.* **1,** 215–237.

Hsiung, G. D., Mannini, Anna, and Melnick, J. L. (1958). Plaque assay of measles virus on *Erythrocebus patas* monkey kidney monolayers. *Proc. Soc. exp. Biol. Med. N.Y.* **98,** 68–70.

Huang, A. S. and Baltimore, D. (1970). Defective viral particles and viral disease processes. *Nature, Lond.* **226,** 325–327.

Hutchins, G. and Janeway, C. A. (1947). Observations on the relationship of measles and remissions in the nephrotic syndrome. *Amer. J. dis. Child.* **73,** 242–243.

Huth, E. (1963). Nephrose—Heilungen nach Masern. *Arch. F. Kinderheilk.* (*Stutt.*) **168,** 178–196.

Imagawa, D. T. (1968). Relationships among measles, canine distemper and rinderpest viruses. *Progr. med. Virol.* **10,** 160–193.

Imagawa, D. T. and Adams, J. M. (1958). Propagation of measles virus in suckling mice. *Proc. Soc. exp. Biol. Med.* **98,** 567–569.

Imagawa, D. T., Yoshimori, M., Wright, S. W. and Adams, J. M. (1954). Serum neutralization of distemper virus in chick embryos. *Proc. Soc. exp. Biol. Med.* **87,** 2–5.

Imagawa, D. T., Goret, P. and Adams, J. M. (1960). Immunological relationships of measles, distemper and rinderpest viruses. *Proc. nat. Acad. Sci. U.S.A.* **46,** 1119–1123.

Iwasaki, Y. and Koprowski, H. (1974). Cell to cell transmission of virus in the central nervous system. 1. Subacute sclerosing panencephalitis. *Lab. Invest.* **31,** 187–196.

Jabbour, J. T., Roane, J. A. and Sever, J. L. (1969). Studies of delayed dermal hypersensitivity in patients with subacute sclerosing panencephalitis. *Neurology,* **19,** 929–931.

Janda, Z., Norrby, E. and Marusyk, H. (1971). Neurotropism of measles virus variants in hamsters. *J. infect. Dis.* **124,** 553–564.

Jensen, M. K. (1968). Lymphocyte transformation in multiple sclerosis. *Acta neurol. scand.,* **44,** 200–206.

Johnson, K. P. and Byington, D. P. (1971). Subacute sclerosing panencephalitis (SSPE) agent in hamsters. 1. Acute giant cell encephalitis in newborn animals. *Exp. mol. Path.,* **15,** 373–379.

Jong, J. G. de (1965). The survival of measles virus in air, in relation to the epidemiology of measles. *Arch. ges. Virusforsch.* **16,** 97–102.

Jong, J. G. de, and Winkler, K. C. (1964). Survival of measles virus in air. *Nature, Lond.* **20** , 1054–2055.

Joseph, B. S. and Oldstone, M. B. A. (1974). Antibody-induced re-distribution of measles virus antigens on the cell surface. *J. Immunol.,* **113,** 1205–1209.

Joseph, B. S. and Oldstone, M. B. A. (1975). Immunologic injury in measles virus infection. II. Suppression of immune injury through antigenic modulation. *J. exp. Med.* **142,** 864–876.

Joseph, B. S., Cooper, N. R. and Oldstone, M. B. A. (1975a). Immunologic injury of cultured cells infected with measles virus. I. Role of IgG antibody and the alternative complement pathway. *J. exp. Med.* **141,** 761–774.

Joseph, B. S., Lampert, P. W. and Oldstone, M. B. A. (1975b). Replication and persistence of measles virus in defined sub-populations of human leukocytes. *J. Virol.* **16,** 1638–1649.

Kadowaki, J., Nihira, M. and Nakao, T. (1970). Reduction of phytohaemagglutinin-induced lymphocyte transformation in patients with measles. *Pediatrics,* **45,** 508.

Kallman, F., Adams, J. M., Williams, R. C. and Imagawa, D. T. (1959). Fine structure of cellular inclusions in measles virus infections. *J. bioph. biochem. Cytol.* **6,** 379–382.

Kalliomäki, J. L. and Halonen, P. (1972). Antibody levels to parainfluenza, herpies simplex, varicella-zostev, cytomegalo virus, and measles virus in patients with connective tissue diseases. *Ann. rheum. Dis.* **31,** 192–195.

Kaluza, G., Scholtissek, C. and Rott, R. (1972). Inhibition of the multiplication of enveloped RNA-viruses by glucosamine and 2-deoxy-d-glucose. *J. gen. Virol.* **14,** 251–259.

Karaki, T. (1965a). Studies on the measles virus in tissue culture. II. Cytochemical and fluorescent antibody studies on the growth of measles virus in FL cells. *J. Jap. Ass. infec. Dis.,* **38,** 445–456.

Karaki, T. (1965b). Studies on measles virus in various tissue cultures. I. Agar cell-suspension plaque assay of measles virus in stable cell lines. *J. Jap. Ass. infec. Dis.,* **38,** 357–366.

Karaseva, I. A., Khozinsky, V. I., Chumakov, M. P., Andreeva, L. S., Petelin, L.S., Fridman, A. Ya., Rostovtseva, R. I., Smirnov. Yu, K. and Graevskaya, N. A. (1974). Antibodies for measles virus in patients with multiple sclerosis. *Tr. Inst. Poliovirus Entset.* **22,** 78–83.

Karelitz, S. and Markham, F. S. (1962). Immunity after modified measles. *Amer. J. dis. Child.,* **103,** 682–687.

Karelitz, S., Schluederberg, A., Kanchanavatee, P., Arai., M. and Acs, H. (1965). Killed measles vaccine prior to seven months of age followed by attenuated live virus vaccine at nine months. *Arch. ges. Virusforsch.* **16,** 343–346.

Karsner, H. T. and Meyers, A. E. (1913). Giant cell pneumonia. *Arch. int. Med.* **11,** 534–541.

Karzon, D. T. (1955). Studies on a neutralizing antibody against canine distemper virus found in man. *Pediatrics,* **16,** 809–818.

Karzon, D. T. (1962). Measles virus. *Ann. N.Y. Acad. Sci.* **101,** 527–539.

Katow, S., Shishido, A., Kobune, K. and Uchida, N. (1973). Growth of measles virus in nervous tissues. *Jap. J. med. Sci. Biol.* **26,** 197–211.

Katz, M. and Koprowski, H. (1973). The significance of failure to isolate infectious viruses in cases of subacute sclerosing panencephalitis. *Arch. ges. Virusforsch.* **41,** 390–393.

Katz, M., Rorke, L. B., Masland, W. S., Koprowski, H. and Tucker, S. H. (1968). Transmission of an encephalitogenic agent from brains of patients with subacute sclerosing panencephalitis to ferrets. *New Engl. J. Med.* **279.** 793–798.

Katz, M., Oyanagi, S. and Koprowski, H. (1969). Subacute sclerosing panencephalitis: Structures resembling myxovirus nucleocapsids in cells cultured from brains. *Nature, Lond.* **222,** 888–890.

Katz, M., Rorke, L. B., Masland, W. S., Barbanti-Brodano, G. and Koprowski, H. (1970a). Subacute sclerosing panencephalitis: Isolation of a virus encephalitogenic for ferrets. *J. infect. Dis.* **121,** 188–195.

Katz, M., ter Meulen, Y., Barbanti-Brodano, G. and Koprowski, H. (1970b). Differentiation in the susceptibility of tissue cultures to subacute sclerosing panencephalitis (SSPE) viruses and measles virus. *Bact. Proc.* 198.

Katz, S. L. (1965). Immunization with live attenuated measles virus vaccines: Five years' experience. *Arch. ges. Virusforsch.* **16,** 222–230.

Katz, S. L. and Enders, J. F. (1959). Immunization of children with a live attenuated measles vaccine. *A.M.A. J. dis. Child.* **85,** 605.

Katz, S. L., Milovanovic, M. V. and Enders, J. F. (1958). Propagation of measles virus in cultures of chick embryo cells. *Proc. Soc. exp. Biol. Med.* **97,** 23–29.

Katz, S. L., Enders, J. F. and Holloway, A. (1962). Use of Edmonston attenuated measles strain. *Amer. J. dis. Child.* **103,** 340–344.

Kenny, M. T. and Schell, K. (1975). Microassay of measles and mumps virus and antibody in Vero cells. *J. Biol Standards,* **3,** 291–306.

Khozinskii, V. I., Chumakov, M. P., Karmysheva, V. Ya., Karaseva, I. A., Atanadze, S. N., Lwonovich, A. L., Kogan, E. I., Tsuker, M. B., Mizinova, I. N., and Lebedeva, I. R. (1974). Experiment of virological and serological study of persisting virus neuro-infection in subacute sclerosing panencephalitis. *Tr. Inst. Poliovirus. Entset.* **22,** 70–77.

Kibler, R. and ter Meulen, V. (1975). Antibody-mediated cytotoxicity after measles virus infection. *J. Immunol.* **114,** 93–98.

Kibler, R., Deller, A. and ter Meulen, V. (1974). Cytotoxic antibody activity in measles and subacute sclerosing panencephalitis (SSPE) infection. *Med. Microbiol. Immunol.,* **160,** 179–190.

Kiley, M. P. and Payne, F. E. (1974a). Replication of measles virus: Continued synthesis of nucleocapsid RNA and increased synthesis of mRNA in the presence of cycloheximide. *J. Virol.,* **14,** 758–764.

Kiley, M. P. and Payne, F. E. (1974b). Evidence of precursors of defective measles virus. *Med. Microbiol. Immunol.* **160,** 91–97.

Kiley, M. P. and Payne, F. E. (1975). Ribonucleic acid synthesis in cells infected with wild- or SSPE-strain measles virus. *In:* "Negative Strand Viruses" (R. D. Barry and B. W. J. Mahy, eds), pp. 531–541. Academic Press, New York and London.

Kiley, M. P., Gray, R. H. and Payne, F. E. (1974). Replication of measles virus: Distinct species of short nucleocapsids in cytoplasmic extracts of infected cells. *J. Virol.* **13,** 721–728.

Kingsbury, D. W. (1970). Replication and functions of myxovirus ribonucleic acid. *Progr. med. Virol.,* **12,** 49–77.

Kingsbury, D. W. (1972). Paramyxovirus replication. *Curr. top. Microbiol. Immunol.* **59,** 1–33.

Kingsbury, D. W. (1974). The molecular biology of paramyxoviruses. *Med. Microbiol. Immunol.* **160,** 74–83.

Kingsbury, D. W., Portner, A. and Darlington, R. W. (1970). Properties of incomplete Sendai virions and sub-genomic viral| RNAs. *Virology,* **42,** 851–871.

Klajman, A., Sternbach, M., Ranon, L., Drucker, M., Geminder, D. and Sadan, N. (1973). Impaired delayed hypersensitivity in subacute sclerosing panencephalitis. *Acta pediat., scand.* **62,** 523–526.

Klenk, H.-D. and Choppin, P. W. (1969). Chemical composition of the parainfluenza virus SV5. *Virology,* **38,** 155–157.

Knight, P., Duff, R. and Rapp, R. (1972). Latency of human measles virus in hamster cells. *J. Virol.* **10,** 995–1001.

Knight, P., Duff, R., Glaser, R. and Rapp, F. (1973–74). Characteristics of the release of measles virus from latently infected cells after co-cultivation with BS-C-1 cells. *Intervirology,* **2,** 287–298.

Knowles, M. and Saunders, M. (1970). Lymphocyte stimulation with measles antigen in multiple sclerosis. *Neurology,* **20,** 700–702.

Kohn, A. (1962). Haemadsorption by measles syncytia. *Nature, Lond.* **193,** 1088–1089.

Kohn, A. and Yassky, D. (1962). Growth of measles virus in KB cells. *Virology,* **17,** 157–163.

Kohno, S., Kohase, M. and Suganuma, M. (1968). Growth of measles virus in a mouse derived established cell line L cells. *Jap. J. med. Sci. Biol.* **21.** 301–311.

Kolakofsky, D. and Bruschi, A. (1973). Molecular weight determination of Sendai RNA by dimethyl sulfoxide gradient sedimentation. *J. Virol.* **11,** 615–620.

Kolakofsky, D., Boy de la Tour, E. and Bruschi, A. (1974). Self-annealing of Sendai virus RNA. *J. Virol.* **14,** 33–39.

Kolar, O. J., Ross, A. T. and Herman, J. T. (1970). Serum and cerebrospinal fluid immunoglobulins in multiple sclerosis. *Neurology,* **20,** 1052–61.

Koprowski, H. (1962). The role of hyperergy in measles encephalitis. *Amer. J. dis. Child.* **103,** 273–278.

Kreth, W. H., Kaeckell, Y. M. and ter Meulen, V. (1974). Cellular immunity in SSPE patients. *Med. Microbiol. Immunol.* **160,** 191–199.

Kreth, W. H., Käckell, M. Y. and ter Meulen, V. (1975). Demonstration of *in vitro* lymphocyte-mediated cytotoxicity against measles virus in SSPE. *J. Immunol.* **114,** 1042–1046.

Krugman, S. (1971). Present status of measles and rubella immunization in the United States: A medical progress report. *J. Pediat.* **78,** 1–16.

Kuwert, E. and Bertrams, J. (1972). Leukocyte iso- and autoantibodies in multiple sclerosis (MS) with special regard to complement-dependent cold-reacting autolymphocytotoxins (CoCoCy). *Euro. Neurol.* **7,** 65–73.

Labowskie, R. J., Edelman, R., Rustigian, R. and Bellanti, J. A., (1974). Studies of cell-mediated immunity to measles virus by *in vitro* lymphocyte-mediated cytotoxicity. *J. infect. Dis.* **129,** 233–239.

Laitenen, O. and Vaheri, A. (1976). Very high measles and rubella virus antibody titres associated with hepatitis, systemic lupus erythematosis and infectious mononucleosis. *Lancet.* **1,** 194–197.

Lamb, R. A., Mahy, B. W. J. and Choppin, P. W. (1976). The synthesis of Sendai virus polypeptides in infected cells. *Virology*, **96**, 116–131.

Lampert, P. W., Joseph, B. S. and Oldstone, M. B. A. (1975). Antibody-induced capping of measles virus antigens on plasma membrane studied by electron microscopy. *J. Virol.* **15**, 1248–1255.

Lazarowitz, S. G., Compans, R. W. and Choppin, P. W. (1971). Influenza virus structural and non-structural proteins in infected cells and their plasma membranes. *Virology*, **46**, 830–843.

Lebon, P., Protat, A. and Molinie, P. (1975). L-arginine elution of measles virus adsorbed on monkey erythrocytes. *Infect. Immun.* **11**, 1407–1408.

Lehrich, J. R., Katz, M., Rorke, L. B., Barbanti-Brodano, G. and Koprowski, H. (1970). Subacute sclerosing panencephalitis. Encephalitis in hamsters produced by viral agents isolated from human brain cells. *Arch. Neurol.* **23**, 97–102.

Lennon, R. G. and Isacson, P. (1967). Delayed dermal hypersensitivity followed killed measles vaccine. Experience in nine-month-old infants. *J. Pediat.* **71**, 525–529.

Lennon, R. G., Isacson, P., Rosales, T., Elsea, W. R., Karzon, D. T. and Winkelstein, W. (1967). Skin tests with measles and poliomyelitis vaccines in recipients of inactivated measles virus vaccine. *J. Amer. med. Assoc.*, **200**, 275–280.

Link, H. (1967). Immunoglobulin G and low molecular weight proteins in human cerebrospinal fluid. Chemical and immunological characterisation with special reference to multiple sclerosis. *Acta neurol. scand.* Vol. 43, Suppl. 28, p. 58.

Link, H., Norrby, E. and Olsson, J. E. (1973). Immunoglobulins and measles antibodies in optic neuritis. *New Engl. J. Med.* **289**, 1103–1107.

Linneman, Jr. C. C. (1973). Measles vaccine: Immunity, re-infection and re-vaccination. *Amer. J. Epidem.* **97**, 365–371.

Lipsey, A. I. and Bolande, R. P. (1967). The exfoliative source of abnormal cells in urine sediment of patients with measles. *Amer. J. dis. Child.* **113**, 677–682.

Lipsey, A. I., Kahn, M. J. and Bolande, R. P. (1967). Pathologic variants of congenital hypogammaglobulinaemia: an analysis of three patients dying of measles. *Pediatrics*, **39**, 659–674.

Lischner, H. W., Sharma, M. and Grover, W. D. (1972). Immunologic abnormalities in subacute sclerosing panencephalitis. *New Engl. J. Med.* **286**, 786–787.

Llanes-Rodas, R. and Liu, C. (1966). Rapid diagnosis of measles from urinary sediments stained with fluorescent antibody. *New Engl. J. Med.* **275**, 516–523.

Lucas, C. J., Brouwer, R. and Feltkamp, T. E. W., ter Veen, J. H. and Van Loghem, J. J. (1972). Measles antibodies in sera from patients with auto-immune diseases. *Lancet i*, 115–116.

M.R.C. Report. Measles Vaccine Committee (1971). Vaccination against measles. Clinical trial of live measles vaccine given alone and live vaccine preceded by killed vaccine. *Practitioner*, **206**, 458–466.

MRC Report. Special Report Series 227. The School Epidemics Committee (1938). Epidemics in schools. An analysis of data collected during the first five years of a statistical enquiry. p. 150. H.M.S.O. Lond.

McCarthy, K. (1959). Measles. *Brit. med. Bull.* **15**, 201–204.

McCarthy, K. (1962). Measles in laboratory hosts and tissue culture systems. *Amer. J. dis. Child.*, **103**, 314–319.

McFarland, H. F. (1974). The effect of measles virus infection on T and B lymphocytes in the mouse. I. Suppression of helper cell activity. *J. Immunol.* **113**, 1978–1983.

McGregor, I. A. (1964). Measles and child mortality in Gambia. *W. Afr. med. J.*, **13**, 251–257.

McKendrick, A. G. (1940). Dynamics of crowd infection. *Edinb. med. J.*, **47**, 117–136.

McLean, D. M., Best, J. M., Smith, P. A., Larke, R. P. B. and McNaughton, G. A. (1966). Viral infections of Toronto children during 1965. II. Measles encephalitis and other complications. *Can. med. Assoc. J.* **94**, 905–910.

McLean, D. M., Kettyls, G. D. M., Hingston, J., Moore, P. S., Paris, R. P. and Rigg, J. M. (1970). Atypical measles following immunization with killed measles vaccine. *Can. med. Assoc. J.* **103**, 743–744.

McNair Scott, T. F. and Bonanno, D. E. (1967). Reactions to live measles virus vaccine in children previously inoculated with killed virus vaccine. *New Engl. J. Med.* **277**, 248–250.

Maeno, K., Yoshida, T., Iinuma, M., Nagai, M., Matsumoto, T. and Asai, J. (1970). Isolation of haemagglutinin and neuraminidase subunits of haemagglutinating virus of Japan. *J. Virol.* **6**, 492–499.

Mahy, B. W. J., Cox, N. J., Armstrong, S. J. and Barry, R. D. (1973). Multiplication of influenza virus in the presence of cordycepin, an inhibitor of cellular RNA synthesis. *Nature, New Biol.* **243**, 172–174.

Majer, M. (1972). Measles virus. *In*: "Strains of human viruses". (M. Majer and S. A. Plotkin, eds), pp. 131–141. S. Karger, Basel.

Makino, S., Sasaki, K., Nazari, F., Nakagawa, M., Nakamura, N. and Kasahara, S. (1970). Cultivation of measles virus in sheep kidney cells. *Jap. J. Microbiol.* **14**, 501–504.

Mannweiler, K. (1965). Ultrastructural examinations of tissue cultures after infection with measles virus. *Arch. ges. Virusforsch.* **16**, 89–96.

Marsden, H. B. (1973). Protracted measles. *Arch. Dis. Child.* **48**, 163.

Martin, S. J. and Brown, F. (1967a). Synthesis of ribonucleic acid in baby hamster kidney cells in the presence of Actinomycin D. *Biochem J.* **105**, 979–986.

Martin, S. J. and Brown, F. (1967b). Effect of virus infection on the stability and synthesis of actinomycin-resistant ribonucleic acid in baby hamster kidney cells. *Biochem. J.* **105**, 987–993.

Marx, P. A., Portner, A. and Kingsbury, D. W. (1974). Sendai virion transcriptase complex: polypeptide composition and inhibition by virion envelope proteins. *J. Virol.* **13**, 107–112.

Mascoli, C. C., Stanfield, L. V. and Phelps, L. N. (1959). Propagation of poliovirus, measles and vaccinia in guinea-pig spleen cell strains. *Science*, **129**, 894–895.

Mastyukova, Yu. N. and Khait, S. L. (1960). The use of tissue cultures for estimation of specific antibodies in antimeasles gamma-globulin. *Vop. Virusol.* **5**, 339–346.

Matsumoto, N. (1966). Studies on measles virus in tissue culture. Electron microscopic study of measles virus infected cells and localization of virus antigen examined by Ferritin-conjugated antibody method. *Bull. Yamaguchi Med. Sch.* **13**, 167–189.

Matumoto, M. (1966). Multiplication of measles virus in cell cultures. *Bact. Rev.* **30**, 152–176.

Matumoto, M., Mutai, M., Ogiwara, H., Nishi, I., Kusano, N. and Aoyama, Y. (1959). Isolement du virus de la rougeole en culture du tissu rénal du Singe. *Compt. Rend. Soc. Franco-Japonaise de Biologie*, **153**, 879–883.

Matumoto, M., Mutai, M., Ogiwara, H. and Nakamura, M. (1961). Prolifération de virus rougeoleux en culture de cellules rénales bovines. *Proc. Soc. Franco-Japonaise de Biologie*, **155**, 1192–95.

Matumoto, M., Saburi, Y., Aoyama, Y. and Mutai, M. (1964). A neurotropic variant of measles virus in suckling mice. *Arch. ges. Virusforsch.* **14**, 683–696.

Matumoto, M., Arita, M. and Oda, M. (1965). Enhancement of measles virus replication by Actinomycin D. *Jap. J. exp. Med.* **35**, 319–329.

Mawhinney, H., Allen, I. V., Beare, J. M., Bridges, J. M., Connolly, J. H., Haire, M., Neill, D. W. and Hobbs, J. R. (1971). Dysgammaglobulinaemia complicated by disseminated measles. *B.M.J.* **2**, 380–381.

Mauler, R. and Hennessen, W. (1965). Virus induced alterations of chromosomes. *Arch. ges. Virusforsch.* **16**, 175–181.

Meadow, S. R., Weller, R. O. and Archibald, R. W. R. (1969). Fatal systemic measles in a child receiving cyclophosphamide for nephrotic syndrome. *Lancet ii*, 876–879.

Melnick, J. L., Benyesh-Melnick, M., Smith, K. O. and Rapp, F. (1965). Recent developments in detection of hidden virus infections. *Perspect. Virol.* **4**, 72–108.

Mellman, W. J. and Wetton, R. (1963). Depression of the tuberculin reaction by attenuated measles virus vaccine. *J. Lab. clin. Med.* **61**, 453–458.

Menna, J. H., Collins, A. R. and Flanagan, T. D. (1975a). Characterization of an *in vitro* persistent-state measles virus infection: Establishment and virological characterization of the BGM/MV cell line. *Infect. Immun.* **11**, 152–158.

Menna, J. H., Collins, A. R. and Flanagan, T. D. (1975b). Characterization of an *in vitro* persistent-state measles virus infection: Species characterization and interference in the BGM/MV cell line. *Infect. Immun.* **11**, 159–163.

Meyer, H. M., Brooks, B. E., Douglas, R. D. and Rogers, N. G. (1962). Potency testing of live measles vaccine. *Amer. J. dis. Child.* **103**, 457–459.

Mihatsch, M. J., Ohnacker, H., Just, M. and Nars, P. W. (1972). Lethal measles giant cell pneumonia after live measles vaccination in a case of thymic alymphoplasia Gitlin. *Helv. paediat. Acta*, **27**, 143–146.

Millar, J. H. D., Fraser, K. B., Haire, M., Connolly, J. H., Shirodaria, P. V. and Hadden, D. S. (1971). Immunoglobulin M specific for measles and mumps in multiple sclerosis. *B.M.J.* **2**, 378–380.

Miller, D. L. (1964). Frequency of complications of measles, 1963. *B.M.J.* **2**, 75–78.

Milovanovic, M. V., Enders, J. F. and Mitus, A. (1957). Cultivation of measles virus in human amnion cells in developing chick embryo. *Proc. Soc. exp. Biol.* **95**, 120–127.

Minagawa, T. (1971a). Studies on the persistent infection with measles virus in HeLa cells. I. Clonal analysis of cells of carrier cultures. *Jap. J. Microbiol.* **15**, 325–331.

Minagawa, T. (1971b). Studies on the persistent infection with measles virus in HeLa cells. II. The properties of carried virus. *Jap. J. Microbiol.* **15**, 333–340.

Minagawa, T. and Yamada, M. (1971). Studies on the persistent infection with measles virus in HeLa cells. III. Immunolysis of cells in carrier state by anti-measles sera. *Jap. J. Microbiol.* **15**, 341–350.

Mirchamsy, H. and Rapp, F. (1969). Role of interferon in replication of virulent and attenuated strains of measles virus. *J. gen. Virol.* **4**, 513–522.

Mirchamsy, H., Razavi, J. and Ahurai, P. (1972). Pathogenesis of vaccine strains of measles virus in suckling hamsters. *Acta virol.* **16**, 77–79.

Mitchell, A. G., Nelson, W. E. and Le Blanc, T. J. (1935). Studies in immunity: V. Effect of acute disease on the reaction of the skin to tuberculin. *Amer. J. dis. Child.* **49**, 695–702.

Mitus, Anna, Enders, J. F., Craig, J. M. and Holloway, A. (1959). Persistence of measles virus and depression of antibody formation in patients with giant-cell pneumonia after measles. *New. Engl. J. Med.*, **261**, 882–889.

Mitus, A., Holloway, A., Evans, A. E. and Enders, J. F. (1962). Attenuated measles vaccine in children with acute leukemia. *Amer. J. dis. Child.* **103**, 413–418.

Mitus, A., Enders, J. F., Edsall, G. and Holloway, A. (1965). Measles in children with malignancy problems and prevention. *Arch. ges. Virusforsch.* **16**, 331–337.

Morley, D. C. (1962). Measles in Nigeria. *Amer. J. dis. Child.* **103**, 230–233.

Morley, D. C. (1969). Severe measles in the tropics—I. *B.M.J.* **1**, 297–300.

Morley, D. C., Martin, W. J. and Allen, I. (1967). Measles in East and Central Africa. *E. Afr. med. J.* **44**, 497–508.

Mottet, N. K. and Szanton, V. (1961). Exfoliated measles giant cells in nasal secretions. *Archs. Path.* **72**, 434–437.

Moulias, R. L., Reinert, Ph. and Goust, J. M. (1971). Immunologic abnormalities in subacute sclerosing panencephalitis. *New. Engl. J. Med.* **285**, 1090.

Mountcastle, W. E., Compans, R. W., Caliguri, L. A. and Choppin, P. W. (1970). Nucleocapsid protein subunits of Simian virus 5, Newcastle disease virus and Sendai virus. *J. Virol.*, **7**, 747–752.

Munyer, T. P., Mangi, R. J., Dolan, T. and Kantor, F. S. (1975). Depressed lymphocyte function after measles-mumps-rubella vaccination. *J. infect. Dis.* **132**, 75–78.

Musser, S. J. and Slater, E. A. (1962). Measles virus growth in canine renal cell cultures. *Amer. J. dis. Child.* **103**, 476–481.

Mutai, M. (1959). Isolation and identification of measles virus. *Jap. J. exp. Med.* **29**, 283–295.

Nader, P. R., Horwitz, M. S. and Rousseau, J. (1968). Atypical exanthem following exposure to natural measles: Eleven cases in children previously inoculated with killed vaccine *J. Pediat.*, **72**, 22–28.

Nakai, M. and Imagawa, D. T. (1969). Electron microscopy of measles virus replication. *J. Virol.* **3**, 187–189.

Nakai, T., Shand, F. L. and Howatson, A. F. (1969). Development of measles virus *in vitro*. *Virology*, **38**, 50–67.

Nalbant, J. P. (1937). The effect of contagious diseases on pulmonary tuberculosis and on the tuberculin reaction in children. *Amer. Rev. Tuberc.* **36**, 773–777.

Nelson, J. D., Sandusky, G. and Peck, F. B. (1966). Measles skin test and serologic response to intradermal measles antigen. *J.A.M.A.* **198**, 653–654.

Neurath, A. R. and Norrby, E. C. J. (1965). Further elucidation of the nature of the measles virus hemolysin based on inactivation studies. *Arch. ges. Virusforsch.*, **15**, 651–658.

Nicolle, C. (1931). La maladie de jeune age des chiens est transmissible expérimentalement à l'homme sous forme inapparantes. Portée de cette constatation. *Arch. inst. Pasteur. Tunis.*, **20**, 321–323.

Nichols, W. W. (1963). Relationships of viruses, chromosomes and carcinogenesis. *Hereditas*, **50**, 53–80.

Nichols, W. W., Levan, A., Hall, B. and Östergren, G. (1962). Measles-associated chromosome breakage. Preliminary communication. *Hereditas*, **48**, 367–370.

Noort, S. van den and Stjernholm, R. L. (1971). Lymphotoxic activity in multiple sclerosis. *Neurology*, **21**, 783–789.

Nordal, J. J., Frøland, S. S., Vandvik, B. and Norrby, E. (1975). Measles-virus-induced migration inhibition of human leucocytes: an immunologically un-specific phenomenon. *Lancet ii*, 1266–1267.

Norrby, E. (1962a). Hemagglutination by measles virus. I. The production of hemagglutinin in tissue culture and the influence of different conditions on the hemagglutinating system. *Arch. ges. Virusforsch.*, **12**, 153–163.

Norrby, E. (1962b). Hemagglutination by measles virus. II. Properties of the hemagglutinin and of the receptors on the erythrocytes. *Arch. ges. Virusforsch.*, **12**, 164–172.

Norrby, E. (1963). Hemagglutination by measles virus. III. Identification of two different hemagglutinins. *Virology*, **19**, 147–157.

Norrby, E. (1964). Separation of measles virus components by equilibrium centrifugation in CsCl gradients. I. Crude and Tween and ether-treated concentrated tissue culture material. *Arch.ges. Virusforsch.*, **14**, 306–318.

Norrby, E. (1967). A carrier cell line of measles virus in Lu 106 cells. *Arch. ges. Virusforsch.*, **20**, 215–224.

Norrby, E. (1971). The effect of a carbobenzoxy tripeptide on the biological activities of measles virus. *Virology*, **44**, 599–608.

Norrby, E. (1972). Intracellular accumulation of measles virus nucleocapsid and envelope antigens. *Microbios.* **5**, 31–40.

Norrby, E. C. J. and Falksveden, L. G. (1964). Some general properties of the measles virus hemolysin. *Arch. ges. Virusforschung.* **14**, 474–486.

Norrby, E. and Gollmar, Y. (1972). Appearance and persistence of antibodies against different virus components after regular measles infections. *Infect. Immun.* **6**, 240–247.

Norrby, E. and Gollmar, Y. (1975). Identification of measles virus-specific haemolysis-inhibiting antibodies separate from haemagglutination-inhibiting antibodies. *Infect. Immun.* **11**, 231–239.

Norrby, E. and Hammarskjöld, B. (1972). Structural components of measles virus. *Microbios.* **5**, 17–29.

Norrby, E. C. J. and Hammarskjöld, B. (1974). A comparison between virion RNA of measles virus and some other paramyxoviruses. *Med. Microbiol. Immunol.* **160**, 99–104.

Norrby, E. C. J. and Magnusson, P. (1965). Some morphological characteristics of the internal component of measles virus. *Arch. ges. Virusforsch.* **17**, 443–447.

Norrby, E., Friding, B., Rockborn, G. and Gard, S. (1964). The ultrastructure of canine distemper virus. *Arch. ges. Virusforsch.* **13**, 335–344.

Norrby, E., Lagercrantz, R. and Gard, S. (1966). Measles vaccination. VI. Serological and clinical follow-up analysis 18 months after a booster injection. *Acta paed. scand.* **55**, 457–462.

Norrby, E., Chiarini, A. and Marusyk, H. (1970). Measles virus variants: Intracellular appearance and biological characteristics of virus products. *In*: "The Biology of Large RNA Viruses" (R. D. Barry and B. W. J. Mahy, eds). pp. 141–153. Academic Press, New York and London.

Norrby, E., Link, H., Olsson, J. E., Panelius, M., Salmi, A. and Vandvik, B. (1974). Comparison of antibodies against different viruses in cerebrospinal fluid and serum samples from patients with multiple sclerosis. *Infect. Immun.* **10**, 688–694.

Norrby, E., Enders-Ruckle, G. and ter Meulen, V. (1975). Differences in appearrance of antibodies to structural components of measles virus after immunization with inactivated and live virus. *J. infect. Dis.*, **132**, 262–269.

Notermans, S. L. H., Tijl, W. F. J., Willems, F. T. C. and Sloof, J. L. (1973). Experimentally-induced sub-acute sclerosing panencephalitis in young dogs. *Neurology*, **23**, 543–553.

Numazaki, Y. and Karzon, D. T. (1966). Density separable fractions during growth of measles virus. *J. Immunol.* **97**, 458–469.

O'Brien, T. C., Albrecht, P., Tauroso, N. M. and Burns. G. R. (1972). Properties of a measles virus neuropathic for Rhesus monkeys. *Arch. ges. Virusforsch.* **39**, 228–239.

Oddo, F. G. and Sinatra, A. (1961). Studi sulla riproduzione del virus morbilloso in colture di tessuti in vitro. *Revista d'ell Instituto Scerotrapico Italiano*, **36**, 65–89.

Oddo, F. G., Flaccomio, R. and Sinatra, A. (1961). "Giant-cell" and "strand-forming" cytopathic effect of measles virus lines conditioned by serial propagation with diluted or concentrated inoculum. *Virology*, **13**, 550–553.

Oddo, F. G., Chiarini, A. and Sinatra, A. (1967). On the hemagglutinating and hemolytic activity of measles virus variants. *Arch. ges. Virusforsch.* **22**, 35–42.

Offner, H., Ammitzbøll, T., Clausen-J., Fog, T., Hyllested, K. and Einstein, E. (1974a). Immune response of lymphocytes from patients with multiple sclerosis to phytohaemagglutinin, basic protein of myelin and measles antigens. *Acta. neurol. scand.* **50**, 373–381.

Offner, H., Konat, G. and Clausen, J. (1974b). Effect of phyto-haemagglutinin, basic protein and measles antigen on myo-(2-^{3}H)-inositol incorporation into phosphatidylinositol of lymphocytes from patients with multiple sclerosis. *Acta. neurol. scand.* **50**, 791–800.

Ojala, A. (1947). On changes in the cerebrospinal fluid during measles. *Ann. med. Intern. Fenn.* **36**, 321–331.

Okuno, Y., Sugai, T., Fujita, T., Yamamura, T., Toyoshima, K., Takahashi, M., Nakamura, K. and Kunita, N. (1960). Studies on the prophylaxis of measles with attenuated living virus. II. Cultivation of measles virus isolated by tissue culture in developing chick egg. *Biken's J.*, **3**, 107–113.

Orvell, C. and Norrby, E. (1974). Further studies on the immunologic relationship among measles, distemper and rinderpest viruses. *J. Immunol.*, **113**, 1850–1858.

Osunkoya, B. O., Cooke, A. R., Ayeni, O. and Adejumo, T. A. (1974a). Studies on leukocyte cultures in measles. I. Lymphocyte transformation and giant cell formation in leukocyte cultures from clinical cases of measles. *Arch. ges. Virusforsch.* **44**, 313–322.

Osunkoya, B. O., Adeleye, G. I., Adejumo, T. A. and Salimonu, L. S. (1974b). Studies on leukocyte cultures in measles. II. Detection of measles virus antigen in human leukocytes. *Arch. ges. Virusforsch.* **44**, 323–329.

Oyanagai, S., ter Meulen, V., Müller, D., Katz, M. and Koprowski, H. (1970). Electron microscopic observations in subacute sclerosing panencephalitis brain cell cultures: their correlation with cytochemical and immunocytological findings. *J. Virol.*, **6**, 370–379.

Oyanagi, S., ter Meulen, V., Katz, M. and Koprowski, H. (1971). Comparison of subacute sclerosing panencephalitis and measles viruses: an electron microscope study. *J. Virol.*, **7**, 176–187.

Palacios, O. (1965). Cytochemical and fluorescent antibody studies on the growth of measles virus in tissue culture. *Arch. ges. Virusforsch.* **16**, 83–88.

Palm, C. R. and Black, F. L. (1961). A comparison of canine distemper and measles viruses. *Proc. Soc. exp. Biol.* **107**, 588–590.

Palmer, D. L., Minard, B. J. and Cawley, L. P. (1976). IgG subgroups in cerebrospinal fluid in multiple sclerosis. *New Engl. J. Med.* **294**, 447–448.

Panelius, M. and Salmi, A. A. (1973). Association of measles antibody activity with electrophoretic fractions of CSF in a patient with multiple sclerosis. *Acta neurol. scand.* **49**, 266–268.

Panum, P. L. (1847). Beobachtungen über das Maserncontagium. *Virchow's Arch.* **1**, 492–512.

Papp, K. (1937). Fixation du virus morbilleux aux leucocytes du sang dès la période d'incubation du maladie. *Bull. Acad. Nat. Med. (Paris).* **117**, 46–51.

Papp, K. (1956). Expériences prouvant que la voie d'infection de la rougeole est la contamination de la muquese conjonctivale. *Revue. Immun. (Paris).* **20**, 27–36.

Parfanovich, M. I. and Sokolov, N. N. (1965). Interaction of nucleic acids and antigens in tissue culture cells doubly infected with measles and tick-borne encephalitis viruses, as revealed by fluorescence microscopy. *Acta. virol.* **9**, 352–357.

Parfanovich, M. I., Sokolov, N. N., Fadeyeva, L. L., Berezina, O. N. and Zhdanov, V. M. (1966). Peculiarities and site of synthesis of viral and cellular ribonucleic acids in measles virus-infected cells. *Acta. virol.* **10** 393–399.

Parfanovich, M., Hammarskjöld, B. and Norrby, E. (1971). Synthesis of virus-specific RNA in cells infected with two different variants of measles virus. *Arch. ges. Virusforsch.* **35**, 38–44.

Parker, J. C., Klintworth, G. K., Graham, D. G. and Griffith, J. F. (1970). Uncommon morphologic features in subacute sclerosing panencephalitis (SSPE). *Amer. J. Path.* **61**, 275–292.

Payne, F. E. and Baublis, J. V. (1971). Measles virus and subacute sclerosing panencephalitis. *Perspectives in Virology*, **7**, 179–195.

Payne, F. E. and Baublis, J. V. (1973). Decreased reactivity of SSPE strains of measles virus with antibody. *J. infect. Dis.* **127**, 505–511.

Payne, F. E., Baublis, J. V. and Itabashi, H. H. (1969). Isolation of measles virus from cell cultures of brain from a patient with subacute sclerosing panencephalitis. *New Engl. J. Med.* **281**, 585–589.

Peebles, T. C. (1967). Distribution of virus in blood components during viraemia of measles. *Arch. ges. Virusforsch.* **22**, 43–47.

Périer, O., Vanderhaegen, J. J. and Pelc, S. (1967). Subacute sclerosing leuco-encephalitis. *Acta neuropathol.* **8**, 362–380.

Périer, O., Thiry, L., Vanderhaegen, J. J. and Pelc, S. (1968). Attempts at experimental transmission and electron microscopic observations in subacute sclerosing panencephalitis. *Neurology*, **18**, 138–143.

Periés, J. R. and Chany, C. (1960). Activité hémagglutinante et hémolytique du virus morbilleux. *Compt. rend. Acad. Sci.*, **251**, 820–821.

Periés, J. R. and Chany, C. (1961). Mécanisme de l'action hémagglutinante des cultures de virus morbilleux. *Compt. rend. Acad. Sci.* **252**, 2956–2957.

Periés, J. R. and Chany, C. (1962). Studies on measles viral hemagglutination. *Proc. Soc. exp. Biol. Med.* **110**, 477–482.

Petermann, M. L. and Pavlovec, A. (1966). The subunits and structural ribonucleic acids of Jensen sarcoma ribosomes. *Biochem. biophys. Acta.* **114**, 264–276.

Phillips, L. A. and Bussell, R. H. (1973). Buoyant density of canine distemper virus. *Arch. ges. Virusforsch.* **41**, 310–318.

Phillips, P. E. and Christian, C. L. (1969). Virologic studies in systemic lupus erythematosus (SLE). *Arthritis Rheum.* **12**, 688–689.

Phillips, P. E. and Christian, C. L. (1970). Myxovirus antibody increases in human connective tissue disease. *Science*, **168**, 982–984.

Pilcher, J. D. (1935). Local skin reactions in measles and scarlet fever in relation to the intracutaneous tuberculin reaction. *Am. Rev. Tuberc.* **3**, 568–575.

Pinkerton, H., Smiley, W. L. and Anderson, W. A. D. (1945). Giant cell pneumonia with inclusions. A lesion common to Hecht's disease, distemper and measles. *Amer. J. Path.* **21**, 1–23.

Pirquet, C. von. (1908). Das Verhalten der kutanen Tuberculinreaktion während der Masern. *Deutsch. med. Wschr.* **34**, 1297–1300.

Plowright, W. (1962). Rinderpest virus. *Ann. N.Y. Acad. Sci.* **101**, 548–563.

Plowright, W. and Ferris, R. D. (1959). Studies with rinderpest virus in tissue culture. I. Growth and cytopathogenicity. *J. comp. Path.* **69**, 152–172.

Popov, V. F., Parkhomov, I. I. and Semenkova, L. I. (1973). Post-vaccination immunity in vaccinees with live measles vaccine. *Vop. Virus.* **18**, 455–458.

Portner, A. and Bussell, R. H. (1973). Measles virus ribonucleic acid and protein synthesis: Effects of 6-azauridine and cycloheximide on viral replication. *J. Virol.* **11**, 46–53.

Portner, A. and Kingsbury, D. W. (1970). Complementary RNAs in paramyxovirions and paramyxovirus-infected cells. *Nature, Lond.* **228**, 1196–1197.

Poste, G., Reeve, P., Alexander, D. J. and Terry, G. (1972). Effect of plasma membrane lipid composition on cellular susceptibility to virus-induced cell fusion. *J. gen. Virol.* **17**, 133–136.

Preble, O. T. and Youngner, J. S. (1975). Temperature-sensitive mutant viruses and the etiology of chronic and inapparent infections. *J. infect. Dis.* **131**, 467–473.

Radyisch, N. S. and Zakharchenko, E. M. (1972). The effect of virulent and vaccine strains of measles virus on the chromosomal apparatus of the cell. *Vop. Virus.* **17**, 611–616.

Raine, C. S., Feldman, L. A., Sheppard, R. D. and Bornstein, M. B. (1969). Ultrastructure of measles virus in cultures of hamster cerebellum. *J. Virol.* 4, 169–181.

Raine, C. S., Feldman, L. A., Sheppard, R. D. and Bornstein, M. B. (1971). Ultrastructural study of long-term measles infection in cultures of hamster dorsal-root ganglion. *J. Virol.* 8, 318–329.

Raine, C. S., Feldman, L. A., Sheppard, R. D. and Bornstein, M. B. (1973). Subacute sclerosing panencephalitis virus in cultures of organized central nervous tissue. *Lab. Invest.* 28, 627–640.

Raine, C. S., Feldman, L. A., Sheppard, R. D., Barbosa, L. H. and Bornstein, M. B. (1974a). Subacute sclerosing panencephalitis virus. Observations on a neuroadapted and non-neuroadapted strain in organotypic central nervous system cultures. *Lab. Invest.* 31, 42–53.

Raine, C. S., Byington, D. P. and Johnson, K. P. (1974b). Experimental subacute sclerosing panencephalitis in the hamster. Ultrastructure of the chronic disease. *Lab. Invest.* 31, 355–368.

Raine, C. S., Byington, D. P. and Johnson, K. P. (1975). Subacute sclerosing panencephalitis in the hamster: Ultrastructure of the acute disease in newborn and weanlings. *Lab. Invest.* 33, 108–116.

Rake, G. and Shaffer, M. F. (1939). Propagation of the agent of measles in the fertile hen's egg. *Nature, Lond.* 144, 672–673.

Rapp, F. (1960). Observations of measles virus infection of human cells. III. Correlation of properties of clones of HEp_{-2} cells with their susceptibility to infection. *Virology*, 10, 86–96.

Rapp, F. (1964). Plaque differentiation and replication of virulent and attenuated strains of measles virus. *J. Bacteriol.* 88, 1448–1458.

Rapp, F., Seligman, S. J., Jaross, L. B. and Gordon, I. (1959). Quantitative determination of infectious units of measles virus by counts of immunofluorescent foci. *Proc. Soc. exp. Biol.* 101, 289–294.

Rapp, F., Gordon, I. and Baker, R. F. (1960). Observations of measles virus infection of cultured human cells. I. A study of development and spread of virus antigen by means of immunofluorescence. *J. biophys. biochem. Cytol.* 7, 43–48.

Rauh, L. W. and Schmidt, R. (1965). Measles immunization with killed virus vaccine. *Amer. J. dis. Child.* 109, 232–237.

Rebel, G., Fontangues, R. and Colobert, L. (1962). Nature lipidique des substances responsable de l'activitité hémolytique de myxovirrus parainfluenzae I (virus Sendai). *Annls. Inst. Pasteur.* 102, 137–152.

Reissig, M., Black, F. L. and Melnick, J. L. (1956). Formation of multinucleated giant cells in measles virus infected cultures deprived of glutamine. *Virology*, 2, 836–838.

Rima, B. and Martin, S. J. (1977). Persistent infection of tissue culture cells by RNA viruses. *Med. Microbiol. Immunol.* 162, 89–118.

Rima, B., Davidson, B. and Martin, S. J. (1976). The role of defective interfering particles in persistent infections of Vero cells by measles virus. *J. gen. Virol.* 35, 89–97.

Rima, B. K., Gould, E. and Martin, S. J. (1977). Variations in proteins synthesized during the establishment of persistent infection with measles virus. *In* "Negative strand viruses and the host cell". (R. D. Barry and B. W. J. Mahy eds.), Academic Press. London and New York. In press.

Roberts, G. B. S. and Bain, A. D. (1958). The pathology of measles. *J. Path. Bact.* **76**, 111–118.

Roberts, J. A. (1965). A study of the antigenic relationship between human measles virus and canine distemper virus. *J. Immunol.* **94**, 622–628.

Robinson, W. S. (1970). Self-annealing of subgroup 2 myxovirus RNAs. *Nature, Lond.* **225**, 944–945.

Robinson, W. S. (1971). Sendai virus RNA synthesis and nucleocapsid formation in the presence of cycloheximide. *Virology*, **44**, 494–502.

Rød, T., Haug, K. W. and Ulstrup, J. C. (1970). Atypical measles after vaccination with killed vaccine. *Scand. J. infect. Dis.* **2**, 161–165.

Romano, N. and Scarlata, G. (1973). Amino acid requirements of measles virus in HeLa cells. *Arch. ges. Virusforsch.*, **43**, 359–366.

Rosanoff, E. I. (1961). Hemagglutination and hemadsorption of measles virus. *Proc. Soc. exp. Biol. Med.* **106**, 563–567.

Rozina, E. E. (1971). Morphological criteria of neurovirulence of viruses in viral vaccine production. *Acta virol.* **15**, 95–101.

Ruckle, G. (1957). Studies with measles virus. *J. Immunol.* **78**, 330–340.

Ruckle, G. (1958). Studies with the monkey-intranuclear-inclusion-agent (MINIA) and foamy agent derived from spontaneously degenerating monkey kidney cultures. I. Isolation and tissue culture behaviour of the agents and identification of Minia as closely related to measles virus. *Arch. ges. Virusforsch.* **8**, 138–166.

Ruckle, G. and Rogers, K. D. (1957). Studies with measles virus. II. Isolation of virus and immunologic studies in persons who have had the natural disease. *J. Immunol.* **78**, 341–355.

Ruckle-Enders, G. (1962). Comparative studies of monkey and human measles virus strains. *Amer. J. dis. Child.* **103**, 297–307.

Rustigian, R. (1962). A carrier state in HeLa cells with measles virus (Edmonston strain) apparently associated with non-infectious virus. *Virology*, **16**, 101–104.

Rustigian, R. (1966a). Persistent infection of cells in culture by measles virus. I. Development and characteristics of HeLa sublines persistently infected with complete virus. *J. Bact.*, **92**, 1792–1804.

Rustigian, R. (1966b). Persistent infection of cells in culture by measles virus. II. Effect of measles antibody on persistently infected HeLa sublines and recovery of a HeLa clonal line persistently infected with incomplete virus. *J. Bact.*, **92**, 1805–1811.

Saburi, Y. and Matumoto, M. (1966). Assay of measles virus haemolysin and its antibody. *Arch. ges. Virusforsch.* **17**, 29–41.

Salmi, A., Gollmar, Y., Norrby, E. and Panelius, M. (1973). Antibodies against three different structural components of measles virus in patients with multiple sclerosis, their siblings and matched controls. *Acta path. microbiol. scand.*, **81B**, 627–634.

Saunders, M., Knowles, M., Chambers, M. E., Caspary, E. A., Gardner-Medwin, D. and Walker, P. (1969). Cellular and humoral responses to measles in subacute sclerosing panencephalitis. *Lancet i,* 72–74.

Savage, F. M. A. (1967). A year of measles. *Med. J. Zambia.* 67–77.

Schluederberg, A. E. (1962). Separation of measles virus particles in density gradients. *Amer. J. dis. Child.* **103,** 291–296.

Schluederberg, A. (1965). Immune globulins in human viral infections. *Nature, Lond.* **205,** 1232–1233.

Schleuderberg, A. (1971). Measles virus RNA. *Biochem. biophys. Res. Commun.* **42,** 1012–1015.

Schluederberg, A. and Chavanich, S. (1974). The role of the nucleus in measles virus replication. *Med. Microbiol. Immunol.* **160,** 85–90.

Schluederberg, A. and Nakamura, M. (1967). A salt-dependent hemagglutinating particle from measles-infected cells. *Virology,* **33,** 297–306.

Schluederberg, A. E. and Roizman, B. (1962). Separation of multiple antigenic components of measles virus by equilibrium sedimentation in cesium chloride. *Virology,* **16,** 80–83.

Schluederberg, A., Williams, C. A. and Black, F. L. (1972). Inhibition of measles virus replication and RNA synthesis by actinomycin D. *Biochem. biophys. Res. Commun.,* **48,** 657–661.

Schluederberg, A., Chavanich, S., Lipman, N. B. and Carter, C. (1974). Comparative molecular weight estimates of measles and subacute sclerosing panencephalitis virus structural polypeptides by simultaneous electrophoresis in acrylamide gel slabs. *Biochem. biophys. Res. Commun.* **58,** 647–651.

Schneck, S. A. (1968). Vaccination with measles and central nervous system disease. *Neurology,* **18,** 79–82.

Schneck, S. A., Fulginiti, V. and Leestma, J. (1967). Measles virus and panencephalitis. *Lancet i,* 1381.

Schumacher, H. P. and Albrecht, P. (1970). Optimal conditions for isolation of a neurotropic measles virus from brain tissue. *Proc. Soc. exp. Biol. Med.* **134,** 396–402.

Schumacher, H. P., Albrecht, P. and Tauraso, N. M. (1972a). The effect of altered immune reactivity on experimental measles encephalitis in rats. *Arch. ges. Virusforsch.,* **37,** 218–229.

Schumacher, H. P., Albrecht, P. and Tauraso, N. M. (1972b). Markers for measles virus. Tissue culture properties. *Arch. ges. Virusforsch.,* **36,** 296–310.

Schwarz, A. J. F. (1964). Immunization against measles: Development and evaluation of a highly attenuated live measles vaccine. *Ann. paediat.* **202,** 241–252.

Schwarz, A. J. F. and Zirbel, L. W. (1959). Propagation of measles virus in non-primate tissue culture. I. Propagation in bovine kidney tissue culture. *Proc. Soc. exp. Biol.* **102,** 711–714.

Seligman, S. J. and Rapp, F. (1959). A variant of measles virus in which giant cell formation appears to be genetically determined. *Virology,* **9,** 143–145.

Sell, K. W., Thurman, G. B., Ahmed, A. and Strong, D. M. (1973). Plasma and spinal-fluid blocking factor in SSPE. *New Engl. J. Med.* **288**, 215–216.

Sergiev, P. G., Ryazantseva, N. E. and Shroit, I. G. (1960). The dynamics of pathological processes in experimental measles in monkeys. *Acta. virol.* **4**, 265–273.

Sergiev, P. G., Shroit, I. G., Chelysheva, K. M., Smirnova, E. V., Kuksova, M. I., Shevtsova, Z. V., Levenshtam, M. A., Chernomordick, A. E., Kozlyuk, A., Stromova, G. I., Manjko, T. G., Yefimov, E. F., Shamrayeva, S. A. and Voskresenskaya, G. S. (1966). Materials on measles pathogenesis and vaccinal process. *Acta virol.* **10**, 430–439.

Sever, J. L. and Kurtzke, J. F. (1969). Delayed dermal hypersensitivity to measles and mumps antigens among multiple sclerosis and control patients. *Neurology*, **19**, 113–115.

Shaver, D. N., Bussel, R. H. and Barron, A. L. (1964). Comparative cytopathology of canine distemper and measles viruses in ferret kidney cell cultures. *Arch. ges. Virusforsch.* **14**, 487–498.

Shingu, M. and Nakagawa, Y. (1960). Studies on the measles virus, the isolation of measles virus on the HeLa cells, and immunological and morphological properties of the isolated agents. *Kurume med. J.* **7**, 82–88.

Shirodaria, P. V., Dermott, E. and Gould, E. A. (1976). Some characteristics of salt-dependent haemagglutinating measles viruses. *J. gen. Virol.* **33**, 107–115.

Shishido, A., Yamanouchi, K., Hikita, M., Sato, T., Fukuda, A. and Kobune, F. (1967). Development of a cell culture system susceptible to measles, canine distemper and rinderpest viruses. *Arch. ges. Virusforsch.* **22**, 364–380.

Shishido, A., Katow, S., Kobune, K. and Sato, T. A. (1973). Growth of measles virus in nervous tissues. I. Neurotropic properties of measles virus in newborn hamsters. *Jap. J. med. Sci. Biol.* **26**, 103–118.

Simpson, R. W. and Iinuma, M. (1975). Recovery of infectious proviral DNA from mammalian cells infected with respiratory syncytial virus. *Proc. nat. Acad. Sci. U.S.A.* **72**, 3230–3234.

Smith, K. O., Gehle, W. D. and Sanford, B. A. (1974). Evidence for chronic viral infections in human arteries. *Proc. Soc. exp. Biol. Med.* **147**, 357–360.

Smithwick, E. M. and Berkovich, S. (1966). *In vitro* suppression of the lymphocyte response to tuberculin by live measles virus. *Proc. Soc. exp. Biol. Med.*, **123**, 276–278.

Smorodintsev, A. A., Boychuk, L. M., Shikina, E. S., Meshalova, V. N., Taros, L. Y., Aminova, M. G., Revenok, N. D. and Safarov, D. I. (1965). Prevention of measles by use of live vaccines in the U.S.S.R. *Arch. ges. Virusforsch.* **16**, 284–293.

Soper, H. E. (1929). The interpretation of periodicity in disease prevalence. *J. R. Stat. Soc.*, **92**, 34–61.

Stanwick, T. L. and Kirk, B. E. (1976). Analysis of baby hamster kidney cells persistently infected with lymphocytic choriomeningitis virus. *J. gen. Virol.*, **32**, 361–369.

Stenback, W. A. and Durard, D. P. (1963). Host influence on the density of NDV. *Virology*, **20**, 545–551.

Stjernholm, R. L., Wheelock, E. F. and Van den Noort, S. (1970). A lymphotoxic factor in multiple sclerosis serum. *J. Reticuloendothelial Soc.* **8**, 334–341.

Stokes, J., Reilly, C. M., Bunyak, E. B. and Hilleman, M. R. (1961). Immunologic studies of measles. *Amer. J. Hyg.* **74**, 293–303.

Stone, J. D. (1948). Prevention of virus infection with enzyme of V. cholerae. II. Studies with influenza virus in mice. *Aust. J. exp. Biol. med. Sci.* **26**, 287–297.

Stone, H. O., Kingsbury, D. W. and Darlington, R. W. (1972). Sendai virus-induced transcriptase from infected cells: polypeptides in the transcriptase complex. *J. Virol.*, **10**, 1037–1043.

Sullivan, J., Barry, D. W., Albrecht, P. and Lucas, S. J. (1975). Inhibition of lymphocyte stimulation by measles virus. *J. Immunol.* **114**, 1458–1461.

Suringa, D. W. R., Bank, L. J. and Ackerman, A. B. (1970). Role of measles virus in skin lesions and Koplik's spots. *New Engl. J. Med.* **283**, 1139–1142.

Symposium (1968). Conference on measles virus and subacute sclerosing panencephalitis. Neurology (Minneap,) 18, No. 1, part 2, 1–192. (J. L. Sever and W. Zeman, eds).

Taniguchi, T., Kamahora, J., Kato, S. and Hagiwara, K. (1954a). Pathology in monkeys experimentally infected with measles virus. *Med. J. Osaka Univ.* **5**, 367–396.

Taniguchi, T., Okuno, Y., Kamahora, J., Aoyama, Y. and Kato, S. (1954b). Animal experiments of measles on monkeys. *Med. J. Osaka Univ.* **4**, 381–397.

Tanzer, J., Stoitchkov, Y., Harel, P. and Boiron, M. (1963). Chromosomal abnormalities in measles. *Lancet ii*, 1070–1071.

Tawara, J. (1964). Micromorphological changes in dog kidney cells infected with measles virus. *Virus.* (*Jap.*) **14**, 85–88.

Tawara, J. (1965). Fine structure of filaments in dog kidney cell cultures infected with measles virus. *Virology*, **25**, 322–323.

Tawara, J. T., Goodman, J. R., Imagawa, D. T. and Adams, J. M. (1961). Fine structure of cellular inclusions in experimental measles. *Virology*, **14**, 410–416.

Tellez-Nagel, I. and Harter, D. H. (1966). Subacute sclerosing leukoencephalitis: Ultrastructure of intranuclear and intracytoplasmic inclusions. *Science*, **154**, 899–901.

ter Meulen, V. and Martin, S. J. (1976). Genesis and maintenance of a persistent infection by canine distemper virus. *J. gen. Virol.*, **32**, 431–440.

ter Meulen, V., Müller, D., Katz, M., Käckell, Y. and Joppich, G. (1970). Immunohistological, microscopical and neurochemical studies on encephalitides. IV. Subacute sclerosing (progressive) panencephalitis. Histochemical and immunohistological findings in tissue cultures derived from SSPE brain biopsies. *Acta neuropath.* **15**, 1–10.

ter Meulen, V., Katz, M., Käckell, Y. M., Barbanti-Brodano, G., Koprowski, H. and Lennette, E. H. (1972a). Subacute sclerosing panencephalitis: *In vitro* characterization of viruses isolated from brain cells in culture. *J. infect. Dis.* **126**, 11–17.

ter Meulen, V., Katz, M., and Müller, D. (1972b). Subacute sclerosing panencephalitis. A review. *Curr. top. Microbiol. Immunol.* **57**, 1–38.

ter Meulen, V., Müller, D., Käckell, Y. M., Katz, M. and Meyermann, R. (1972c). Isolation of measles virus in measles encephalitis. *Lancet i*, 1172–1175.

ter Meulen, V., Katz, M. and Käckell, Y. M. (1973). Properties of SSPE virus; tissue culture and animal studies. *Ann. clin. Res.* **5**, 293–297.

Thamdrup, E. (1952). Re-infections with measles: Familial immunity defect. *Acta paediat.* **41**, 267–282.

Thein, P., Mayr, A., ter Meulen, V., Koprowski, H., K[a]ckell, M. Y., Müller. D. and Meyermann, R. (1972). Subacute sclero[s]ing panencephalitis: Transmission of the virus to calves and lambs. *Arch· Neurol.* **27**, 540–548.

Thiry, L., Dachy, A. and Lowenthal, A. (1969). Measles antibodies in patients with various types of measles infection. *Arch. ges. Virusforsch.* **28**, 278–284.

Thomson, D., Connolly, J. H., Underwood, B. O. and Brown, F. (1975). A study of immunoglobulin M antibody to measles, canine distemper and rinderpest viruses in sera of patients with subacute sclerosing panencephalitis. *J. clin. Path.* **28**, 543–546.

Thorne, H. V. and Dermott, E. (1976). Circular and elongated linear forms of measles virus nucleocapsid. *Nature, Lond.* **264**, 473–474.

Thurman, G. B., Ahmed, A., Strong, D. M., Knudsen, R. C., Grace, W. R. and Sell, K. W. (1973). Lymphocyte activation in subacute sclerosing panencephalitis virus and cytomegalovirus infections. *In vitro* stimulation in response to viral-infected cell lines. *J. exp. Med.*, **138**, 839–846.

Thygeson, P. (1959). Ocular viral diseases. *Med. Clins. N. Am.* **43**, 1419–1440.

Tijl, W. F. J., Notermans, S. L. H., Katz, M., ter Meulen, V. and Koprowski, H. (1971). SSPE virus-encephalitis produced experimentally in young dogs. Clinical and electro-encephalographic aspects. *Proc. 13th Int. Cong. Pediatrics.* Vienna. 3, 425–429.

Tischer, I. (1967). Erhöte Empfindlichkeit von neuraminidase-behandelten Affenerythrocyten für den Nachweiss der Haemagglutinations aktivitat von Masernvirus und einigen Adenoviren. *Zentralb. Bakt. Abt. 18. Orig.* **205**, 466–476.

Tikhonova, N. T., Khrometskaya, T. M., Kholchev, N. V., Streltsova, M. P. and Nesterova, T. P. (1973). Study of physico-chemical nature of anti-measles antibody in children who had typical measles. *Vop. Virus.* **18**, 27–32.

Todd, D. and Martin, S. J. (1975). Determination of molecular weight of bovine enterovirus RNA by nuclease digestion. *J. gen. Virol.* **26**, 121–129.

Tokuda, G., Fukusho, K., Morimoto, T. and Watanabe, M. (1963). Studies on rinderpest virus in bovine leukocyte culture. II. Susceptibility of leukocyte culture to the virus. *Nat. Inst. Anim. Hlth. Q. (Jap.)* **3**, 55–63.

Tourtellotte, W. W. (1975). What is multiple sclerosis? laboratory criteria for diagnosis. *In* "Multiple Sclerosis Research". (A. N. Davison, J. H. Humphrey, L. A. Liversedge, W. I. McDonald and J. S. Porterfield, eds), pp. 9–26. H.M.S.O. Lond.

Toyoshima, K., Hata, S., Takahashi, M., Kunita, N., and Okuno, Y. (1959a). Virological studies on measles virus. II. Growth of Toyoshima strain in four established cell lines *Biken's J.* **2**, 313–320.

Toyoshima, K., Takahashi, M., Hata, S., Kunita, N. and Okuno, Y. (1959b). Virological studies on measles virus. I. Isolation of measles virus using FL cells and immunological properties of the isolated agents. *Biken's J.* **2**, 305–312.

Toyoshima, K., Hata, S. and Miki, T. (1960a). Virological studies on measles virus. IV. The effect of active and inactivated measles virus on cultured cells. *Biken's J.* **3**, 281–291.

Toyoshima, K., Hata, S., Takahashi, M., Miki, T. and Okuno, Y. (1960b). Virological studies on measles virus. III. Morphological changes and virus growth in FL cultures. *Biken's J.* **3**, 241–248.

Triger, D. R., Kurtz, J. B., MacCallum, F. O. and Wright, R. (1972). Raised antibody titres to measles and rubella viruses in chronic active hepatitis. *Lancet i*, 665–667.

Triger, D. R., Gamlen, T. R., Paraskeuas, E., Lloyd, R. S. and Wright, R. (1976). Measles antibodies and autoantibodies in autoimmune disorders. *Clin. exp. Immunol.* **24**, 407–414.

Ueda, S., Okuno, Y., Sakamoto, Y., Sangkawibha, N., Tuchinda, P., Ochatanonda, P., Bukkavesa, S., Yamada, Y., Suzuki, K., Tanami, Y., Kusano, F., Hayakawa, Y. and Kurose, T. (1974). Studies on further attenuated live measles vaccine. VIII. Estimated duration of immunity after vaccination without natural infection. *Biken's J.,* **17**, 11–20.

Underwood, G. E. (1959). Studies on measles virus in tissue culture. I. Growth rates in various cells and development of a plaque assay. *J. Immunol.* **83**, 198–205.

Underwood, B. and Brown, F. (1974). Physico-chemical characterisation of rinderpest virus. *Med. Microbiol. Immunol.* **160**, 125–132.

Utermohlen, V. and Zabriskie, J. B. (1973). Suppressed cellular immunity to measles antigen in multiple sclerosis patients. *Lancet ii*, 1147–1148.

Utermohlen, V., Winfield, J. B., Zabriskie, J. B. and Kunkel, H. G. (1974). A depression of cell-mediated immunity to measles antigen in patients with systemic lupus erythematosis. *J. exp. Med.* **139**, 1019–1024.

Valdimarsson, H., Agnarsdottir, G. and Lachmann, P. J. (1975). Measles virus receptor on human T lymphocytes. *Nature, Lond.* **255**, 554–556.

Vandvik, B. and Degre, M. (1975). Measles virus antibodies in serum and cerebrospinal fluid in patients with multiple sclerosis and other neurologic disorders, with special reference to measles antibody synthesis within the central nervous system. *J. neurol. Sci.* **24**, 201–219.

Waksman, B. H., Burnstein, T. and Adams, R. D. (1962). Histologic study of the encephalomyelitis produced in hamsters by a neurotropic strain of measles. *J. Neuropath. exp. Neurol.* **21**, 25–49.

Walker, D. L. (1964). The viral carrier state in animal cultures. *Progr. med. Virol.*, **6**, 111–148.

Warren, J., Nadel, M. K., Slater, E. and Millian, S. J. (1960). The canine distemper—measles complex. I. Immune response of dogs to canine distemper and measles virus. *Amer. J. vet. Res.*, **21**, 111–119.

Warthin, A. S. (1931). Occurrence of numerous large giant cells in the tonsils and pharyngeal mucosa in the prodromal stage of measles. *Archs. Path.* **11**, 864–874.

Waters, D. J. and Bussell, R. H. (1973). Polypeptide composition of measles and canine distemper virus. *Virology*, **55**, 554–557.

Waters, D. J. and Bussell, R. H. (1974). Isolation and comparative study of the nucleocapsids of measles and canine distemper virus from infected cells. *Virology*, **61**, 64–79.

Waters, D. J., Hersh, R. T. and Bussell, R. H. (1972). Isolation and characterization of measles nucleocapsid from infected cells. *Virology*, **48**, 278–281.

Waterson, A. P. (1962). Two kinds of myxovirus. *Nature. Lond.* **193**, 1163–1164.

Waterson, A. P. (1965). Measles virus. *Arch. ges. Virusforsch.* **16**, 57–80.

Waterson, A. P. and Almeida, J. D. (1966). Taxonomic implications of myxovirus. *Nature, Lond.* **210**, 1138–1140.

Waterson, A. P., Cruickshank, J. G., Laurence, G. D. and Kanarek, A. D. (1961). The nature of measles virus. *Virology*, **15**, 379–382.

Waterson, A. P., Rott, R. and Ruckle-Enders, G. (1963). The components of measles virus and their relation to rinderpest and distemper. *Z. Naturforsch.* **18B**, 377–384.

Wear, D. J. and Rapp, F. (1970). Encephalitis in newborn hamsters after intra-cerebral injection of attenuated human measles virus. *Nature, Lond.* **227**, 1347–1348.

Wear, D. J. and Rapp, F. (1971). Latent measles virus infection of the hamster central nervous system. *J. Immunol.* **107**, 1593–1598.

Wear, D. J., Rabin, E. R., Richardson, L. S. and Rapp, F. (1968). Virus replication and ultra structural changes after induction of encephalitis in mice by measles virus. *Exp. mol. Path.* **9**, 405–417.

Weber, M., Canton, P. and Dureux, J. B. (1969). Aspects E.E.G. au cours de la rougeole. *Ann. Med. Nancy*, **8**, 655–660.

Wild, F., Underwood, B. and Brown, F. (1974). Ribonucleic acid synthesis in rinderpest virus infected cells. *Med. Microbiol. Immunol.* **160**, 133–141.

Winston, S. H., Rustigian, R. and Bratt, M. A. (1973). Persistent infection of cells in culture by measles virus. III. Comparison of virus-specific RNA synthesized in primary and persistent infection in HeLa cells. *J. Virol.* **11**, 926–932.

Wright, T. (1957a). Isolation of measles virus. *B.M.J.* **1**, 1241.

Wright, T. (1957b). Cytopathic effect on primate and rodent tissue-cultures of agent isolated from measles patient. *Lancet i*, 669–670.

Yamanouchi, K., Egashira, Y., Uchida, N., Kodama, H., Kobune, F., Hayami, M., Fukuda, A. and Shishido, A. (1970). Giant cell formation in lymphoid tissues of monkeys inoculated with various strains of measles virus. *Jap. J. med. Sci. Biol.* **23**, 131–145.

Yamanouchi, K., Chino, F., Kobune, F., Kodama, H. and Tsurahara, T. (1973). Growth of measles virus in the lymphoid tissues of monkeys. *J. infect. Dis.* **128**, 795–799.

Yamanouchi, K., Chino, F., Kobune, F., Fukuda, A. and Yoshikawa, Y. (1974). Pathogenesis of rinderpest virus infection in rabbits. I. Clinical signs, immune response, histological changes and virus growth patterns. *Infect. Immun.* **19**, 199–205.

Yeh, J. (1973). Characterisation of virus-specific RNAs from subacute sclerosing panencephalitis virus-infected CV-1 cells. *J. Virol.* **12**, 962–968.

Young, L. W., Smith, D. I. and Glasgow, L. A. (1970). Pneumonia of atypical measles. Residual nodular lesions. *Am. J. Roentg.* **110**, 439–448.

Young, N. P. and Ash, R. J. (1970). Polykaryocyte induction by NDV propagated on different hosts. *J. gen. Virol.* **7**, 81–82.

Zabriskie, J. B. (1975). Cell mediated immunity to viral antigens in multiple sclerosis. *In*: "Multiple Sclerosis Research" (A. N. Davison *et al.* eds), pp. 142–150. H.M.S.O. London.

Zhdanov, V. M. (1961). Recent experience with antiviral vaccines. *Ann. rev. Microbiol.* **15**, 297–322.

Zhdanov, V. M. (1975). Integration of viral genomes. *Nature, Lond.* **256**, 471–473.

Zhdanov, V. M. and Parfanovich, M. I. (1974). Integration of measles virus nucleic acid into the cell genome. *Arch. ges. Virusforsch.* **45**, 225–234.

Zweiman, B. (1971). In vitro effects of measles virus on proliferating human lymphocytes. *J. Immunol.* **106**, 1154–1158.

Zweiman, B. (1972). Effect of viable and non-viable measles virus on proliferating human lymphocytes. *Int. Archs. Allergy*, **43**, 600–607.

Zweiman, B. and Miller, M. F. (1974). Effects of non-viable measles on proliferating human lymphocytes. II. Characteristics of the suppressive reaction. *Int. Archs. Allergy.* **46**, 822–823.

Zweiman, B., Pappagianis, D., Maibach, H. and Hildreth, E. A. (1971). Effect of measles immunization on tuberculin hypersensitivity and *in vitro* lymphocyte reactivity. *Int. Archs. Allergy.* **40**, 834–841.

Subject Index